Imagination in Kant's Critical Philosophy

Imagination in Kant's Critical Philosophy

—

Edited by
Michael L. Thompson

DE GRUYTER

ISBN 978-3-11-048165-5
e-ISBN 978-3-11-027465-3

Library of Congress Cataloging-in-Publication Data
A CIP catalog record for this book has been applied for at the Library of Congress

Bibliographic information published by the Deutsche Nationalbibliothek
The Deutsche Nationalbibliothek lists this publication in the Deutsche Nationalbibliografie;
detailed bibliographic data are available in the Internet at http://dnb.dnb.de.

© 2013 Walter de Gruyter GmbH, Berlin/Boston
Printing and binding: Hubert & Co. GmbH & Co. KG, Göttingen
Printed on acid-free paper
Printed in Germany

www.degruyter.com

Contents

Michael Thompson
Introduction

Imagination created the world[1]

Through this passage of an indeterminate
product of the free power of imagination to its
total determination in one and the same act,
that which occurs in my consciousness becomes
an image [*Bild*] and is posited as an image. It
becomes *my* product because I must posit it
through absolute self-activity.[2]

It is little coincidence that immediately following the wide distribution of Kant's philosophy we find a surge of literary and philosophical authors extolling the imagination as imperative to our cognitive functioning. Kant himself demonstrates the necessity and obscurity of this capacity in his famously cryptic passage:

> Synthesis in general, as we shall hereafter see, is the mere result of the power of imagination (*Einbildungskraft*), a blind but indispensable function of the soul, without which we should have no knowledge whatsoever, but of which we are scarcely ever conscious.[3]

Just what the imagination is and how it serves our varying mental activities in Kant presents a formidable challenge to any student of Kant's philosophy. Part of this difficulty is due to conflicting and aporetic doctrines—amongst which one finds the most salient and obscure discussions surround Kant's view of *Einbildungskraft*.

This book aims to recover the lacuna and elucidate this often overlooked faculty in Immanuel Kant's critical philosophy. The primary thesis in this volume is that the complexity and robustness of Kant's metaphysical, epistemological, aesthetic and moral theories cannot be accounted for fully without appeal to the imagination and the products of its activities. By situating the imagination within the entirety of Kant's critical philosophy a story about the imagination and Kant's cognitive architectonic can be told. Due to technical vocabulary, complex-

---

1 Baudalaire, Charles (1962): "La Reine des Facultés" in *Curiosités esthétiques [et] L'Art romantique*. H Lemaitre (Ed.) Paris: Garnier, p. 321.
2 Fichte,Johann Gottlieb (1991) *The Science of Human Knowledge [Wissenschaftslehre]* Heath and Lachs (transl) Cambridge: Cambridge University Press, p.3.
3 Kant, Immanuel (1965): *The Critique of Pure Reason* A78/B103 trans. Norman Kemp Smith. New York: Macmillan & Co, p. 112.

ity of thought and overall intricacy of Kant's philosophical position, isolating any one element of his cognitive apparatus in order to make clear its function, status, role and employment in cognition presents an interpreter with a number of challenges. For example, isolating sensibility from the rest of the cognitive structures e. g. the understanding and reason, and determining its constituent role in knowledge production appears to be nearly impossible, if not entirely so. How can one understand this element without reference to its counterpart, and, furthermore, how can one clearly determine its role in cognition without the contraposing faculty with which it combines in knowledge production? By focusing on one element in Kant's philosophy, one runs the risk of failing to illustrate said element's proper place in Kant's critical philosophy. And yet, one cannot understand Kant's philosophy without providing an analytic of the elements by means of which one can isolate constitutive parts, illustrate their roles and determine them in their interactions. For this volume, I would like to propose that an isolation of one element is not only possible, but also necessary in an interpretation, defense and emendation of Kant's critical works. By focusing on the imagination, one will be able to interpret and defend Kant's critical evaluation of scientific, metaphysical, practical and aesthetic knowledge. The intent here is to focus on the imagination in order to gain greater insight on this "blind but indispensible function".

The Traditional Conversation

One finds Kant's initially substantive discussion of the imagination in a section of his *Critique of Pure Reason* entitled the Transcendental Deduction of the Table of Categories; by Kant's own admission, the section of the book that cost him the most labor. Instrumental in these most critical passages are his discussions of the varying roles the imagination plays in our cognitive processes, but, due to revisions, emendations and a seeming change in doctrine from his 1st Critique (1781, 1787) to his 3rd Critique (*Critique of Judgment* 1790), what Kant's considered view of the imagination is remains unclear and has been largely overlooked. Several authors eschew discussion of this primary faculty, dismissing it as arcana of an obsolete faculty psychology. Even prominent Kant scholars have typically overlooked or marginalized pivotal sections of Kant's works in order to avoid dealing with this issue.

The reactions to Kant's table of categories and his purported deduction of them are as variable as they are numerous. Importantly, the various assessments of Kant's critical exegesis of imagination are even more capricious than the estimates of the deductions themselves. An exhaustive account here might take us too far afield from the discussion of the imagination in Kant, but we can elucidate general trends and objections authors have noted over the years. The most general trend we find in these authors is a harsh critique leveled at what Kant has claimed to achieve in the transcendental deduction. Commonly, Kant is charged with having provided a faculty psychology that explains what processes are in play in judgments, even the a priori grounds by which cognition obtains, but, the criticism typically cites, the faculty psychology does not provide a detailed process of the imagination nor its role in category production, or how synthesis occurs. Notably, Hermann Cohen[4] rejects the deduction of the table of categories, instead preferring to read the Transcendental Analytic in reverse order. Cohen begins with the Analytic of Principles and interprets them as an epistemology of Newtonian physics. By claiming Newtonian physics as an a priori science of the principles of experience, Cohen argues that the Kant's elucidation of the Analytic of Principles provides the principles applied in cognition and believes the table of categories can be deduced therefrom.[5] By demonstrating how knowledge is possible, i.e. the principles applied in judgment, Cohen believes we can deduce the categories without the need to appeal to imagination as a synthetic function. In brief, Cohen argues that by knowing what it is that we call knowledge and how we come to these claims, we can deduce the constitutive half of knowledge found in the understanding. This strategy may be the way Kant actually conceived his critique of reason. It is plausible that Kant presupposed Euclidean geometry as an a priori science, and proceeded to provide a faculty psychology and the principles necessary to affirm this assumption. His presentation, however, proceeds in a very different manner. What Cohen fails to realize is that Newtonian physics cannot be an a priori natural science, because the principles found in Newton are derived from experience, hence have an empirical condition and cannot be pure a priori, although they may be a priori.[6] Laws of gravitation and momentum may seem to be universal and necessary

---

4 Founder of the Marburg school of neo-Kantianism circa 1902, whose adherents include Paul Natorp, Ernst Cassirer, and eventually many logical positivists through the influence of Rudolph Carnap. For further discussion see Michael Friedman's (2000) *The Parting of the Ways Chicago: Open Court* pp. 25–26.

5 Cf. Cohen, Hermann (1918) *Kant's Theorie der Erfahrung* 3rd ed. Berlin: B. Cassirer pp. 345–346..

6 In the introduction to CPR Kant makes a distinction between pure a priori and a priori. The former indicates the universality and necessity required prior to any experience. The latter can be construed as universal and necessary, but are dependent upon empirical conditions. As an example of the latter, Kant cites that with proper understanding of structural engineering, one need not undermine the foundations of a house to know that if one does, the roof collapses. One can know a priori what will happen, but this a priori knowledge is dependent upon the em-

for the objects of experience, but the legitimacy they boast always has its sources in abstraction from empirical examples. Indeed, they may govern empirical objects as far as we have seen them demonstrated, but they are proven inductively and hence do not possess the a priority necessary to be a pure natural science.

P.F. Strawson continues in the neo-Kantian, analytic tradition by arguing for a failure of the transcendental deduction.[7] Strawson interprets the purpose of the deduction to be a more modern use of the term "deduction", a deduction of the categories in a sense that Kant does not himself seem to endorse. For Strawson a deduction needs to provide a genesis for the categories themselves from axiomatic principles, and in so doing must explain the principles of synthesis at work via the imagination (or understanding, as he would prefer). Strawson argues that objective validity can only be achieved if the very conceptual architecture, the categories themselves, can be demonstrated in their universality and necessity and their genesis and employment. This most certainly is not what Kant provides, and Strawson deems Kant's exercise as a complete failure.

In a more sympathetic vein, Henry Allison attempts to redress Strawson's accusations and to defend Kant against undue interpretation.[8] He points out what Kant means by objective validity, noting Strawson's misunderstanding, and he attempts to ward off the pronouncement of complete failure. However, Allison himself admits Kant's lack when it comes to an explanation of the table of categories.[9] And while Allison admits the conspicuous lack of a deduction in terms of the origin and genesis of the categories, he mitigates Strawson's critique further by citing the schematism as the illustration of the application of the categories to intuition. While Allison attempts to present a defense of Kant's transcendental idealism, he continues the tradition of reading the 1st *Critique* primarily as a treatise on epistemology. The defense of transcendental idealism is made by distinguishing between empirical and transcendental idealisms and focuses on the epistemic conditions that Kant offers to argue for the latter.

Martin Heidegger will oppose the trend to read Kant's work exclusively as epistemology, claiming Kant has performed an invaluable service explicating

---

pirical conditions set forward by engineering. The former indicates knowledge prior to any empirical conditions. Cf. CPR A8/B12 Cp. A21/B35.

7 Strawson, P.F. (1966) *The Bounds of Sense* New York: Routledge p. 117.

8 Strawson here exemplifies the mid-20th century analytic approach to the deduction. Other authors include H.A. Prichard, Jonathan Bennett and might be characterized as trying to purge the idealism from Kant in an effort to uphold the Copernican insight Kant displayed, but to save Kant from himself.

9 Allison, Henry (1983) *Kant's Transcendental Idealism* New Haven: Yale University Press p. 170.

the regional ontology of human knowledge.[10] He interprets the doctrine of the transcendental power of imagination as an illustration of *Dasein's* finitude and fundamental orientation to time. The source of pure concepts of human cognition is to be found in this very orientation to time. Yet, a Heideggerian reading of Kant presents its own difficulties. His analysis of Kant's use of time in structuring the categories and their application remains faithful to Kant's intended explicit statements, but space appears to have been lost in Heidegger's analysis. Furthermore, Heidegger accuses Kant of not having gone far enough. According to this reading, Kant may have seen the ontological implications of his own work, implications Heidegger will make explicit in terms of his own fundamental ontology; but, Heidegger accuses, Kant failed to move beyond delimitations of human cognition, and by not doing so failed to draw the philosophical connection between his epistemology and fundamental ontology. Yet Kant was neither concerned with nor familiar with this subsequent development of fundamental ontology and thus Heidegger has been accused of reading too much into Kant's employment of time. That is to say, Heidegger reads too much of his own philosophy into that of Kant.

Recently, Beatrice Longuenesse attempts to reformulate the question of the source of the categories. Rather than looking exclusively to the Deduction of the Principles, she follows Kant's own suggestion that the table of categories finds its sources in the transcendental table of logical judgments.[11] In the so-called metaphysical deduction of the categories, Kant himself makes explicit the connection between the table of judgments and the table of categories whilst making use of imagination, but what he fails to provide is what this connection might be. By exploring the table of judgments and the arrived body of logic during Kant's time, namely Aristotelian syllogistic logic, Longuenesse attempts 1) to recreate how logic and subsumption work in this logical system, in order to demonstrate the a priority of the categories, 2) to demonstrate their necessity in order to make judgments and 3) their origin itself. What remains unclear is the origin of the categories. Her line of argumentation achieves the first and second of the

---

10 Heidegger represents the competing school of neo-Kantianism in early 20th Century Germany. In contrast to the logico-epistemic reading found in Cohen's Marburg school, Heidegger and the so-called Southwest school, founded by Wilhelm Windelband in Heidelberg and continued by Heinrich Rickert in Freiburg, insist on the distinction between math, logic and the table of categories. Within the Southwest school, Heidegger's particular approach is to interpret Kant's work as a pre-formulation of Heidegger's own project of fundamental ontology. Cf. Friedman, Michael (2000) *The Parting of the Ways Chicago: Open Court* p. 26 – 33.
11 Longuenesse, Beatrice (1998) *Kant and the Capacity to Judge* C.T. Wolfe (translator) Princeton: Princeton University Press p. 5.

three stated goals, but it remains questionable whether she achieves the final task. Longuenesse demonstrates how syllogistic judgments work, and even illustrates how the categories are employed in the categorical premises of syllogisms, thus connecting major and minor premises and showing how universal concepts are necessary in order to make particular judgments. The categories can be proven necessary for judgments and their role in doing so can even be illustrated, but what Longuenesse does not seem to describe is how the categories are supposed to arise from the judgments themselves through the use of imagination. One suggestion is that the table of logical judgments itself represents the necessary means by which any judgment can be made. And if we must judge according to these forms, there must be some concept employed in order to make the possibility of general predication possible in a categorical, hypothetical or disjunctive statement. Thus she believes that from the necessity of judgments arise the need and list of the categories Kant has provided. One difficulty with the interpretation centers around what Kant considers being the origin of the categories. Such an explanation may indeed demonstrate how they are employed in judgment and the necessity of them in use, but it speaks very little toward the source from which categories arise, that is, prior to application in use.

Common to all these interpretations, except Heidegger's, is a focus on the B-edition deduction. Recently, however, a new interest in the imagination, its role in the Transcendental Deductions, what part it plays as a source for category genesis and what role it plays in the larger cognitive sphere has resurfaced. In a departure from the traditional analysis of the deductions and its subsequent reliance upon the B-edition, these recent authors discuss the crucial passages from both the A and B-edition deductions as well as relevant passages from his other critical works.

An Emended Discussion

For Kant, one chief concern is the connection between the two stems of human knowledge. One of his central foci is the question: how are concepts and intuitions brought together to form knowledge? With two stems, Kant finds himself in need of a common root. Imagination, it is suggested, might very well be this radical connection. Hence questions concerning the status, function and rules of operation by which the imagination exercises its task are critical in understanding Kant's entire philosophy. This reported intent might imply a narrow confine to the 1st Critique, but, I believe, such an approach is short-sighted. Kant's employment of the imagination is not merely limited to epistemic claims concerning the connection of human thinking/judging to objects. The imagination figures prominently in all aspects of connecting sensibility with the understanding in judgments, whether of metaphysical, epistemic, moral or aesthetic. When it comes to determining the appropriateness of applying a priori categories to the deliverances of the senses, judgment is the central issue, and, in the 3rd Critique, judgment is the focus of concern. Therefore, concern with the 3rd Critique and its explicit treatment of the imagination is also in order. Furthermore, a look to the 2nd Critique is in store to determine the role of imagination, if any, in moral judgments.

Rather than approaching this topic through regular means—by examining the arguments exclusively found in the B-edition Transcendental Deduction of the 1st Critique, an approach most Anglo-American Kant scholars pursue—the approach here is to cross-examine the imagination in Kant's revisions and several of his works. Treatments of the imagination in Kant's corpus, however, one finds to be remarkably incomplete. More often than not, imagination is discussed in context of Kant's 3rd Critique and the analysis of aesthetic judgments. But it is precisely with the aesthetic that Kant begins his critical enterprise. Therefore, this volume proposes to examine the imagination not only in context of Kant's 3rd Critique, but also in terms of the Transcendental Aesthetic of the 1st Critique and its connection to the Transcendental Analytic. Looking into Kant's moral writings, attention must be made to what use we find of the imagination in the 2nd Critique and the *Grundlegung*. The general thesis of this work is that Kant's use of the imagination is well-informed and radical, and can be found throughout his entire critical philosophy. By employing the imagination as a critical capacity, Kant transforms the imagination from the specious and mistrusted faculty of tradition into a necessary element of human thinking.

More often than not, Kant and his employment of imagination are relegated to marginal treatment or, worse, isolated to a passing footnote. For those authors that do treat the imagination in Kant more extensively, the focus is isolated on one or perhaps a few texts.[12] Very few authors attempt to bring a continuous and comprehensive examination of imagination in Kant's critical corpus. Sarah Gibbon's work, *Kant's Theory of the Imagination*, is the only text that attempts an integrationist account of the imagination in all three critiques. Conspicuously lacking, however, is much connection to the pre-critical period, the *Anthropology* and the *Opus Postumum*.[13] Moreover, by emphasizing "the possibility of cognition from the point of view of the judging subject"[14] and the media-

---

12 Makkreel, Crowther, Kneller etc. Plus a host of papers on imagination and perception.
13 With the exception of Kant's *Inaugural Dissertation*.
14 Gibbons, Sarah (1994): *Kant's Theory of Imagination*.Oxford: Clarendon Press, p. 6.

tional role of imagination, Gibbons misses a more radical origin of the resources of cognition, that is from the imagination itself, and the possibility of addressing the source, genesis of the categories as well as the common root of Kant's two stems of human knowledge.[15] No volume exists that attempts to integrate a comprehensive and radical view of the imagination and its employment in Kant's writings. This volume presents an overture to this integrationist approach, which focuses on the "critical" Kant while attending to selected passages from the "pre-critical" and "post-critical" periods as well. Such a work is fraught with difficulties, some of which I would like to list and briefly explain here, in order to orient the interpretive strategy as well as demonstrate the often protracted fight in Kant scholarship.

Interpretive Issues

The difficulty of this inquiry is compounded by several factors. Not only are Kant's primary texts often inexact, obscure and inconsistent, but secondary authors discussing the imagination in Kant are in radical disagreement concerning how the inquiry should be approached. Scholarship on this issue is divided as well as divisive. While there are well-established translations, well-rehearsed arguments and well-defined doctrines of Kantian philosophy, methodological and conceptual disagreements have relegated the field to certain fiefdoms, which, once certain claims are made, are bitterly defended. And, as forays are made into other interpretive lands, exploratory, invasive raids are made in attempt to expand empires. These empires, much like feudal lands, are bequeathed to trusted vassals and inheritors of the realms. At the outset, there are vested interests about which school of Kantian interpretation one follows.

Much as we find in other research areas, Kant scholars disagree on conceptualizations, methodology and specificity. Depending on whether one pursues the Marburg, Southwest or Anglo-American schools of thought, disparate interpretations and infighting occur on issues aesthetic, metaphysical, epistemic, moral and now even environmental. One oversight in the establishment of these feudal properties is a holistic approach. Much current scholarship confines itself to the well-documented "critical period," roughly 1781–1894. And even within this specific time frame, may authors wish to isolate and ignore passages, often constrained to a single work, that prove fruitful grounds to the continuing

---

15 A theme certainly understood by some of his contemporaries and many of his students, and one which is capitalized on by his immediate inheritors, the German Idealists.

discussions. Inherent in this narrow approach is a marginalization of the system-aticity Kant described as his critical philosophy. Epistemic, moral and aesthetic cognitions all follow along an architectonic that follows from fundamental struc-tures of our human capacities. By limiting themselves to a single work, or even single sections within a given work, many interpreters fail to recognize Kant's own commitment to a single architectonic. Moreover, the "pre-critical" period and the late writings of an academic in retirement, the *Opus Postumum*, may lend themselves to continuing this holistic picture. By attending to passages and overtures made, one that might present a coherent narrative to Kant's life and works rather than the disparate story commonly told.[16] An integrationist ap-proach is, however, fraught with peril of its own. With internal inconsistency, evolution of ideas, different versions of the same texts, difficulties surrounding legitimacy of late texts and seeming ravings at the end of his life, attempting to provide an account of the entirety of Kant comes across as fool-hardy. Inter-pretations of Kant's employment of the imagination finds itself with an abun-dance of source material and yet no cohesion.

There are, furthermore, additional difficulties surrounding such an ap-proach. The first, and perhaps most disconcerting, problem with this proposed study is the possibility to present an inaccurate, superficial and incomplete ac-count of Kant's imagination, thus misrepresenting what such a faculty plays in his thought. Because this volume attempts to trace the employment of imagina-tion in Kant's philosophy, attending to the use, modification, and perhaps even development of such a theme in Kant's corpus, the materials available are nu-merous and often seemingly contradictory. The purpose of this study is not to overlook, dismiss, marginalize or explain away what might appear as conflicts or contradictions. The purpose is to attempt a unifying theme that can ground Kant's philosophical use of imagination and to see its place in the overarching issues of his work. Addressing the seeming inconsistencies and attempting to find a grounding by which Kant can maintain his arguments is the task this work sets out to accomplish. The task is admittedly a large one, but one which I believe attainable, if one attends to the over-riding concern of elaborating the role of imagination in judgments, that is, in the origins of the categories of

---

16 Manfred Kuehn is one author who makes excellent inroads overcoming the common con-ception that Kant breaks completely with a Leibnizian-Wolffian philosophy during the "silent decade" after which he begins his "critical period." Kuehn illustrates trends in Kant's thinking that present developmental connections from the so-called periods of Kant's life. While not a biography, this work follows this lead by drawing connections between different works and periods of Kant's life in order to show the development and importance of imagination in his thought. Cf. Kuehn, Manfred (2001): *Kant: A Biography*. Cambridge: Cambridge University Press.

the understanding and their connection with the deliverances of the senses in the several types of judgments Kant enumerates.

This approach finds sympathy, not only with the pre-critical Kant and his metaphysical inquiries, but with the post-critical period and Kant's concerns with unifying his system. The former, albeit the more rationalistic approach of the Leibniz-Wolffian school, does concern itself with the origins of the contents of the "inner" realm. In these works, Kant explores the basic principles that govern human thinking e. g. the principles of non-contradiction, succession and simultaneity such as those found in the *New Elucidations*. The post-critical period, cited as Kant's works in the years following 1794, finds an attempted summary in the *Opus Postumum* and this work attempts to bring together the insight of the Critiques and scientific exploration of the empirical world; that is, practical application of the insights found in the critical period and the deliverances of the senses found in scientific inquiry. The critical period, it would seem, is bookended by the very concerns of the Critiques themselves. The work accomplished here is to establish a core doctrine of the imagination in the Critiques, that further research into the connectivity of Kant's works may find traction.

A second concern with such a study is the terminological shifts we find throughout Kant's lifetime. Kant's use of imagination found in the pre-critical period is in alignment with the typical use found in the history of philosophy. In *Dreams of a Spirit Seeker*, Kant employs the Latinate *focus imaginarius* to describe the process by which impressions of external bodies produce spatial images available to judgments by the understanding.[17] And while this process is necessary to coordinate "inner" representations with "outer" objects, the opportunity for figments of the fantastical imagination arises. Kant claims it is quite necessary that one "cannot, as long as [one] is awake, fail to distinguish my imaginings, as the figments of my own imagination, from the impressions of the senses."[18] In Kant's own employment of imagination in this work, he subscribes to the general tendency in the history of philosophy to concede the necessity of the imagination, while cautioning his audience to the pernicious nature of fantastical imagination.[19] At this point Kant does glimpse the necessity

---

17 Kant, Immanuel *Dreams of a Spirit Seeker,* in: Academy, 2:345 and 2:347. "Dreams of a Spirit Seeker" in Theoretical *Philosophy 1755–1770* (Cambridge: Cambridge University Press) 1992
18 Ibid.
19 Failing to differentiate between the image-making function of imagination and the fantastical employ of the imagination is to fall victim to "that type of mental disturbance which is called madness, and which, if it is more serious, is derangement." 2:346. In *Dreams*, Kant attributes those visions of shamans, spirit-seekers and, in particular, to Emmanuel Swedenborg, to just such a mistake. Cf. Kuehn (2001), pp. 170–1.

of imagination, without providing much detail in the role it will play in connecting sensibility with understanding. At this early stage in his development, Kant continues the standard historical use of the imagination, one that concedes its employment, but condemns the imagination in its misapplied use. Kant will never truly deviate from this basic position, hence his connection with the history of the imagination. What Kant will develop in his mature writings, however, is insight into the means by which the imagination will perform its role as a liaison, giving the imagination its proper due, while cautioning against its overuse, into inquiries that human reason "is not able to ignore," but which "it also not able to answer."[20]

In the critical period Kant will discuss several different imaginations; the reproductive imagination, the productive imagination, the transcendental imagination, and, it has been argued, even replaces the faculty of sensibility in the 3rd Critique with the term "imagination" itself. In this effort to discuss the imagination, these various uses must be brought into relief, providing distinctions as Kant presents them, but also uniting them under a general use of imagination. The insights found in the critical period are also marked by a shift in linguistic usage. Kant does employ the Latinate "*imaginatio*", but more commonly employs the German term "*Einbildungskraft.*" The shift from Latin to German in his writing coincides with a deeper insight into the formative power of imagination. The shift to his native language and his subsequent philosophical insights may be attributed in part to his newfound critical programmatic, but may also be a shift from the image centered *imaginatio* to a power of creating, building and culture, *Einbildungskraft.* While keeping the image-making function of the historical reproductive imagination, Kant gains new respect for the formative and creative powers of imagination. And even though Kant finds new respect for the imagination in the critical period, he still cautions against its overuse in speculative metaphysics.

Imagination does not figure into Kant's post-critical thought too largely. One explanation for this is that much of his published works are re-figurations of lectures and previously written manuscripts. The attention of these works is often to "scientific" inquiries, notably his *Anthropology* and *Opus Postumum.* What we find in these works is rare mention of the imagination, often in a derogatory tone. However, what insight we find into the imagination is its application in empirical pursuits. After the critical work is accomplished in the three *Critiques*, Kant finds no need to discuss the imagination, but attends to the application of the processes discovered earlier. Following Manfred Kuehn, I would like to

---

20 CPR Avii.

suggest that Kant may develop many of his ideas, but does not deviate too greatly from his overall quest to establish metaphysics as a secure science and to explore the appropriate realms for human inquiry, both scientific and moral.

A third and deep concern for any study is the interpretation of the major thinker the author brings to his analysis. The question of concern is: Just what Kant are you reading? This particular issue has become one aspect of the cottage industry that is Kant scholarship. For authors with overriding epistemic concerns, the 1st Critique is the primary focus and support for argumentation is drawn chiefly from this text. For those interested in moral or aesthetic issues, the texts primarily sought are the 2nd Critique and Groundwork, and 3rd Critique, respectively. Typically, one finds these divisions demarcated by an ocean or channel. Anglo-American interpretations, with their main focus on epistemology, often attempt to separate "the analytic argument" from Kant's transcendental idealism.[21] More European interpretations that focus on aesthetic and moral dimensions often separate themselves from Kant's 1st Critique emphasizing a development or change in Kant's position.[22] When comparing Anglo-American interpretations with those more European, one often finds a sharp contrast between strict analytic approaches that attempt to reconstruct Kant's arguments and evaluate them accordingly and more historical approaches that attempt to contextualize the arguments found in Kant's work. Recently, however, we find overtures to bridge the gap between these two Kants, notably in the works of Beatrice Longuenesse and Hannah Ginsborg.

These two branches of Kant scholarship, while geographically significant, find their radical division in the immediate reaction to Kant's critical works. The European group finds itself charting the historical progression of Kant's ideas through German Idealism and the Southwest school of interpretation. The Anglo-American trend follows a more logical trajectory through the works of Frege and the neo-Kantianism that arose in the early 20th Century through

---

21 Authors such as P.F. Strawson and H.A. Prichard present a "standard" interpretation that purports to demonstrate the incoherence of Kant's project, and yet reserving room to extract the analytic arguments that they deem to be correct in Kant's work. Cf Strawson (1966) *The Bounds of Sense* p. 15–16.

22 This group appears less exclusive in their analysis of Kant. Often short digressions into Kant's 1st Critique and analysis of judgment are afforded by those authors who wish to treat the *Critique of Judgment* properly. Of course, there are exceptions to these generalized statements about Kant scholars.

the Marburg school of interpretation.[23] Moreover, at the heart of the division between interpretive strategies is a conflict concerning which version of the 1st *Cri-*
tique is Kant's more considered view. Noting Kant completely revised several sections, provided an entirely new preface, introduction, and transcendental deduction, along with additions to his refutation of idealism and a, perhaps, radical
and contradictory reformulation of his analogies of experience, the B-edition
contains what some consider to be considerable differences from the A-edition.
The most significant of these changes, so the debate contends, is Kant's rewriting
of the transcendental deduction. This question appears to have become one of
the most divisive, if not the most, in Kant scholarship. The Anglo-American tradition argues that Kant's considered view is the B-edition. After its initial publication, subsequent criticism in the literature, notably the Garve-Feder review,
and reaction, Kant reformulates the heart of his philosophical enterprise, the
transcendental deduction, in order to distinguish himself more clearly from an-
tecedent forms of idealism. In order to distance his transcendental idealism from
the metaphysical or naïve idealism of Berkeley, Kant rewrites the transcendental
deduction and adds a refutation of idealism. The Southwest school of interpretation, broadly the more European interpretation, countenances this argument,
but cites the originality and insightfulness found in the A-edition transcendental
deduction. Such an interpretation argues that the original formulation is the
truer expression of Kant's philosophical position, and that the reformulation is
merely an attempt to allay critics who misunderstood the original, and thus is
Kantian, but not Kant's considered view. The B-edition, they contend, is a reaction to criticism, and perhaps an attempt at popularization, not the advancement
of his ground-breaking philosophical insight. The protracted debate is typically
resolved by favoring one edition over the other and explaining away the discrep-
ancies found between the two by subsuming one under the other.

Such interpretive strategies appear to be a plausible way to resolve the differences between the different versions. But to overlook the insight of one edition
in favor of the other is to tacitly concede that Kant changes his position between

---

23 The Marburg school, founded by Hermann Cohen and closely associated with Paul Natorp
and Ernst Cassirer, favor a reading of Kant that emphasizes interpretations that adhere more
closely to a reading that supports a Fregean, positivistic, logical framework that emphasizes
epistemic concerns over those metaphysical. The Southwest school, founded by Wilhelm Windelband and continued by Henrich Rickert, emphasizes the substantive logic described by Kant
as transcendental logic. Hegel and the German Idealists consider themselves elaborators of this
transcendental logic and emphasize the dialectical, metaphysical content over the formal,
general logic favored by the Marburg school. Cf. Friedman, Michael (2000): *The Parting of the*
Ways. Chicago: Open Court .

1781 and 1787. This is not the approach favored in this volume. Certainly the A-edition of the transcendental deduction has advantages over the B-edition. The attention to detail, the continuity of terminology and the detailed connection and progression from the Transcendental Aesthetic is more pronounced. And yet, the B-edition appears to enlarge the scope, while omitting some of the details found in the A-edition. By locating the insights and elaborating the continuity and coherence between the two editions and Kant's other works, one can, I believe, not only determine the role of imagination in cognition, but also provide insight into the different ways one can putatively employ such a faculty. In addition to the synthetic function of imagination in apprehension, reproduction and recognition of the deliverances of the senses, as found in the A-edition, Kant will also distinguish between intellectual and figurative syntheses in the B-edition. Both versions of the transcendental deduction must be taken into account in order to elaborate the comprehensive scope of imagination in Kant's philosophy. Thus, while I favor the A-edition for its insight and originality, I also concede the advancements made in the B-edition and its attempt to bring the radical insight from the earlier version into discussion with the philosophical conceptualizations of Kant's time. The authors brought together here seem to share this sympathy and attempt to provide justice to the differences in editions in the 1st Critique, as well as considered positions in reconciling the 1st and 3rd Critique.

By pursuing this approach I consider myself aligned more with the Southwest school of Kantian interpretation, highlighted and developed in philosophers such as Martin Heidegger, George Sherover, Martin Weatherston and Dieter Henrich, but also admit the benefit of exploring bracing examinations of Kant's arguments as found in the Anglo-American tradition. Such is the spirit I find in Henry Allison's work *Kant's Transcendental Idealism* and Longuenesse's *Kant and the Capacity to Judge*, a commitment to an explanation and defense of Kant's work, but a commitment to examining Kant's arguments and a willingness to point out when they do not achieve what he believed them to have accomplished.[24] Perhaps, the core of the argument for the radical use of the imagination in Kant's philosophy is just such a critique. Heidegger has pointed out (and the claim has been much discussed) that Kant may have glimpsed the truly remarkable place the imagination occupies in Kant's transcendental arguments, but that he shrank back from the abyss—and I wish to assume just such a

---

24 Allison, Henry (1983): *Kant's Transcendental Idealism*.New Haven, Mass: Yale University Press.

stance.[25] But rather than simply accepting Heidegger's often confusing analysis of imagination, the present authors offer their own: the imagination does occupy a central place in Kant's critical philosophy, but one that is not merely a propaedeutic to *Dasein* analysis. Rather, it is fundamental in all aspects of our cognition. Kant's own transcendental deduction does not provide such an explanation for the pure concepts and, this has been argued, presents a failure of the most critical portion of Kant's work. I concede that what the transcendental deduction provides is not exactly what the name implies, but the work provided in this section is also necessary in order to complete Kant's task in providing such a more straightforward deduction of pure concepts themselves. Kant's deduction is not a failure, as most Anglo-American scholarship suggests, but also does not go far enough, as Heidegger claims.

In light of these difficulties in scope and interpretation, I propose to recognize them here at the outset, and ask indulgence of the reader to address such concerns as they arise between the various authors. Within the analysis of the *Critique of Pure Reason* alone, this last interpretive concern looms large. In attempting to draw connections between Kant's works, terminological and continuity issues arise. These concerns cannot be ameliorated at one single insistence, but only by being faithful and charitable to Kant's own writings, while attempting to critizise, develop and draw the connections implicit in his writings.

Overview of the Essays

In the first essay of this volume, Angelica Nuzzo addresses the status of the imagination in Kant's *Critique of Pure Reason* by focusing his transcendental theory of faculties and the arguments in favor of imagination as a mediator between sensibility and understanding. While denying the meditational role, and the "common root" theory of Heidegger, Nuzzo emphasizes the role of imagination in sensibility and the synthesis that takes place within the manifold, both empirical and a priori. Moreover, Nuzzo determines the active role of imagination in sensibility and illustrates how the standard distinctions between passive sensibility and active understanding are mistaken in light of the role imagination and synthesis play in *Sinnlichkeit*.

The second essay continues to address questions raised in Kant's assessment of the Transcendental Aesthetic, namely mathematics as forms of intuition, and

---

25 Heidegger, Martin (1997): *Kant and the Problem of Metaphysics*. 5th ed. trans. Richard Taft. Bloomington: Indiana University Press, p. 112.

the standard interpretation that sensibility is merely passive and mathematics cannot be informed by imagination. Christian Helmut Wenzel presents an informed discussion of the practice of theoretical mathematics and the processes by which practitioners arrive at new theoretical constructs. By employing both the 1st Critique and the *Critique of Judgment*, Wenzel argues not only that mathematics is a creative and artistic enterprise, but one that must, necessarily be informed by the use of Kant's productive imagination.

Gary Banham's selection discusses imagination at the very heart of the *Critique of Pure Reason, the Transcendental Deduction*. Banham sets out to provide an argument for the transcendental synthesis of imagination as the centre-piece of Kant's transcendental deduction, and as something the interpretation of which was requisite to understand the way Kant's transcendental psychology relates to his transcendental logic. The point of this piece is to uncover the way that the description of transcendental psychology Kant provides gives a distinctive account of the genesis of experience that can be essentially expounded in a way that makes it analytically distinct from the argument concerning the categories.

Sidney Axinn turns his attention to the differences in images, signs, schemata and symbols in Kant's critical philosophy. The connection between imagination and schemata (and mental images for that matter) is well-documented in Kant scholarship, and Axinn addresses what role the imagination might play in symbols, noticeably in moral and practical symbols. Analogizing between symbols employed in mathematics and physics, Axinn argues that beauty is in fact a symbol of morality, and one that is provided by the free play of imagination in both aesthetic and moral judgments.

In *"Functions of Imagination in Kant's Moral Philosophy"*, Bernard Freydberg concentrates on a particularly vexing lacuna in Kant scholarship and imagination. Kant employs the term "imagination" in only a couple of passages in his moral writings. By retreading the functions of imagination in Kant's 1st Critique, Freydberg distills a role of imagination in Kant's moral philosophy, namely as the faculty that schematizes. According to Freydberg, the categorical imperative and its type remain too abstract for direct application, and it is to the imagination that one must look in order to determine what bridges the gap between theoretical constructs and empirical reality. This "primacy of imagination, is essential for any moral action. Included in this is the function of imaginative envisioning closing the gap between the theoretical type, principles for action, and empirical phenomena, actions themselves, by providing an empirical schemata.

Fernando Costa Mattos continues the exploration of imagination in Kant's moral philosophy by providing a "slightly less ambitious perspective" than Freydberg, thus arguing that it is in Kant's postulates of pure practical reason that one finds a role for the imagination. By emphasizing the ideas of God, freedom, the soul, and the highest good in both Kant's theoretical and practical writings, Mattos offers an interpretation of the imagination that is not so much responsible for application of the moral law, but, rather, as an extension of theoretical postulates, that helps guide not merely our actions but the motivating force behind them.

Jane Kneller's contribution to this volume extends her interpretation of Kant's imagination beyond the aesthetic realm and into a more global inspection of imagination and its necessity in all possible experience. Kneller argues against Paul Guyer's interpretation of Kant's relegation of imagination to merely time-determination in favor of a more systemic employment of imagination and imaginative synthesis in regulating the affinity of the manifold. By endorsing Kant's simple definition of imagination, Kneller present a more comprehensive interpretation to illustrate the imagination's employment in all possible forms of experience, theoretical, empirical and aesthetic.

Emily Brady's essay presents the connection between Kant's moral philosophy and the *Critique of Judgment*. In this essay she discuss how imagination operates in more positive ways and show how its activity is intimately tied up with sublime judgment and feeling. More specifically, she argues that in the mathematically sublime, imagination is expanded through attempts to capture the infinite, an activity that can be described in terms of aesthetic freedom, the sublime counterpart, as it were, for the free play of imagination in the beautiful. In the dynamically sublime, we find that imagination functions negatively in being overwhelmed by powerful natural qualities, yet also positively through modes of projection and identification. Imagination's negative and positive functioning is crucial to the feeling of moral freedom which emerges through this second type of sublimity. Through an exploration of the constructive functioning of imagination in the sublime, she extends our understanding of this mental power in Kant's philosophy and to link its activity to different modes of freedom.

In his interpretive essay, "Imagination, Progress and Evolution", Martin Schönfeld provides insight into both Kant scholarship and the fecundity of such pursuits. By emphasizing the differences between scholarly work and evocative thinking (*Geisterbeschwörung*), Schönfeld calls to attention the importance of not merely studying the texts and remaining charitable to such a venerable figure such as Kant, but also to employment of insight and principles found in the works of such a master. Schönfeld finds traction for the insights of Kant and imagination in moral and evolutionary matters, while elaborating how the imagination enlarges our scope of thinking and provides applicability for our moral considerations. By joining considerations of Kant's critical philosophy with tenets of his naturalistic, pre-critical period, Schönfeld provides "a unified account of the power of imagination from general cognition to moral cognition to heuristic progressions and all the way to evolutionary leaps."

Rudolf Makkreel presents the last essay in this collection, and it is fitting that he does so. Not only has he been working on imagination and Kant for multiple decades and thus provides a considered view on the matter, Makkreel also demarcates the influences from Kant's philosophy to Kant's successors. Makkreel combines imagination in Kant over his entire corpus to Wilhelm Dilthey's historical imagination to provide a stunningly original role for imagining the construction, interpretation and orientation of our world. This essay is not only original and powerful in Kant scholarship itself, but is also in the spirit of Kant's imagination. It is absorption, incorporation and projection of the very functions of imagination, and Makkreel illustrates how the imagination provides orientation, interpretation and meaning to our lives.

Angelica Nuzzo

Imaginative Sensibility Understanding, Sensibility, and Imagination in the *Critique of Pure Reason*

The function that Kant assigns to the power or faculty of the imagination (*Einbildungskraft*)[1] in the *Critique of Pure Reason*—and the shifting importance that he attributes to it from the 1781 to the 1787 edition—is uncontroversially recognized as crucial not only for an understanding of fundamental moments of this work such as the thesis of transcendental idealism, the transcendental deduction of the categories, and the schematism, but also for an understanding of the historical role that Kant's philosophy plays at the end of the early modern tradition and at the threshold of Romanticism. The extensive literature on the topic is remarkably consistent in insisting, on the one hand, on the alternatively mediating, intermediary, or even duplicitous position of the imagination, which Kant seems to place *between* sensibility and understanding, the sensible and the intellectual, concept and intuition, ambiguously assigning to it precisely those characters proper to each of them whereby they are distinguished from each other.[2] On the other hand, interpreters are unanimous in underscoring the prominence to which Kant raises this mental power in comparison to early modern epistemologies and psychological theories of the mental faculties—this new prominence being also, at the same time, the crucial Kantian inheritance and inspiration from which many romantic philosophers of the successive era will take their cue.[3] There is, however, an interesting silence around the two connected issues of the very intermediate position that the imagination occupies in the theory of the first *Critique*, and of the mediation that the imagination allegedly effects in the relation between sensibility and understanding, concept and intuition. In other words, *that* the imagination plays, for Kant, a mediating function is generally considered unproblematic, a self-evident claim that apparently does not require further investigation. The interpretive issue seems to regard only the precise way in which the imagination does indeed execute such a function (in

---

[1] For a reflection on Kant's terminology and its development from the pre-critical period on see Makkreel (1990), chapter 1.

[2] As we will see below, the imagination is a "blind" yet "spontaneous" faculty—blindness being the distinctive feature whereby sensibility is set apart from the understanding, spontaneity being the distinctive feature whereby the understanding is opposed to sensibility.

[3] This topic is at the center of Kneller (2007), chapters 1, 6–7 in particular.

the relation, for example, between Transcendental Aesthetic and Analytic, in the second part of the B deduction, in the schematism)—and on this point the interpretations vary as widely as on hardly any other issue raised by this work. By contrast, in this essay I want to draw attention to the former, more general problem. Why is the *imagination*, of all mental powers, allegedly charged with a *mediating* function? That is, why the imagination, and why mediation? And what is, exactly, the mediation or intermediary, bridging task that Kant assigns to the imagination? While post-Kantian philosophy and particularly Hegel have accustomed us (more or less consciously, to be sure) to take mediation in a very definite and by and large accepted sense, it is far less clear to what extent Kant's alleged "dualism" (in whichever way this may be construed) can—and indeed should—tolerate mediation even while seemingly advocating its need (as is explicitly the case in the schematism chapter).[4] In fact, that in Kant's critical philosophy mediation does constitute a crucial problem—and an all but self-evident one at that—is testified by the final, sustained reflection of the *Critique of Judgment* in which the issue of mediation (and, not coincidentally, the role of the imagination) becomes the central theme.[5]

In this essay I address the general problem of the status of the imagination in the transcendental theory of the faculties of the first *Critique* and propose to re-think the issue of the alleged mediating task of the imagination in this work. Briefly put, the problem is the following: given that for Kant two are the irreducible branches of human knowledge, namely, *Sinnlichkeit* and *Verstand*,[6] does the imagination belong to the former or to the latter? Interestingly, interpreters (and perhaps Kant himself) tend to dodge this question although they feel compelled to at least indirectly pose it (but not necessarily to answer it). In any case, as they do not seem too concerned with the uneasiness and wavering that the question produces, they are also not particularly intrigued by it. Thus, in contrast to the prevailing views, this is the problem that I propose to investigate in what follows. What is it, in the theory of the first *Critique* that makes it so difficult to indicate in clear-cut terms whether the imagination is a function of sensibility or an activity of the understanding? Why does the need arise of either qualifying the answer with some restricting condition gesturing to the other side of the al-

---

4 See, for example, the designation of the schema as "vermittelnde Vorstellung" at KrV B177/ A138; the schema is a "Produkt der Einbildungskraft" (B179/A139). Indeed, the post-Kantian theme of "mediation" is deeply connected to the rejection of Kantian "dualism"; the more mediation is admitted into the philosophical vocabulary as a legitimate and even necessary procedure the more dualism is abandoned.

5 See, in general, Nuzzo (2005), chapters 6–8.

6 KrV B29/A15.

ternative or of presenting the imagination as a possible "third" or middle and distinct term between the two, or even, with Heidegger's influential reading, as that very "common although to us unknown root" to which both sensibility and understanding must be brought back, thereby rejecting the alternative altogether?[7] These are relevant questions if only because just to raise them necessarily requires one to revisit the meaning and the motivation of the fundamental distinction on which Kant's entire critical project rests. For, if sensibility and understanding, on the ground of the mediating, bridging role of the imagination, ultimately turn out to be not so separate after all, transcendental philosophy can hardly avoid either relapsing into the early modern gradualism according to which sensible and intellectual representations are not different in kind but only in degree (of clearness and distinctness), or admitting the possibility of an intellect that is itself intuitive (or of an intuition that is itself intellectual). The fact that in both cases the project of the first *Critique* is inexorably undermined gives a sense of how high the stakes are in the question of the status of the imagination in relation to *Sinnlichkeit* and *Verstand*. It is in this connection that we need to ask whether "mediation" here helps solve the problem or rather muddles it—whether it denotes Kant's gesturing toward his philosophical aftermath or is instead a later interpretive imposition on a theory that constitutively shuns any attempt to overcome or undermine dualism. In the latter case, the function of the imagination needs to be characterized differently than through mediation.

Let me briefly anticipate the claim I want to advance in the argument below and the steps in which I shall carry it out. As suggested above, I address the two general and interconnected issues that, I believe, lead us to the very core of Kant's transcendental project. I set out, first, to probe whether the question of the imagination's belonging to either sensibility or the understanding does allow for a clear-cut choice between the two. The answer to this question, which I address by discussing passages from the Transcendental Aesthetic and the B deduction, will occupy this essay for the most part.[8] In discussing this question, however, I indirectly offer a reflection on the issue of the alleged me-

---

7 KrV B29/A15. Heidegger's famous reading (Heidegger, 1973) and Henrich's critique of it (Henrich, 1955) are often discussed in the literature, see among the many references, Kneller (2007, 95 – 97); Makkreel (1990, 21 ff.).

8 In this regard, although I will discuss passages of §24 and §26 of the B deduction, my aim is not directly an interpretation of this chapter of the *Critique* but the limited issue of the role of the imagination in the relation between sensibility and understanding. I have to leave out, for reasons of space, a discussion of the schematism. This, however, will occupy a second essay in the near future.

diating function that Kant assigns to the imagination, and on the implications of such a function for the critical project as a whole. How shall mediation be understood given the imagination's placement in the transcendental theory of the faculties offered by the *Critique of Pure Reason?*

Although, as mentioned above, the literature remains ambiguous on the question of the imagination's belonging either to sensibility or the understanding, the general tendency is to align it with the understanding while making sure to spell out the proviso for this choice. I shall attempt instead the opposite reading. I suggest that the imagination belongs to *Sinnlichkeit* or is a specific function proper to sensibility, advancing this claim in order to measure the novelty of Kant's *transcendental* idea of (human) sensibility. To put this point differently, I contend that imagination is a sensible faculty but I argue that the peculiar imaginative function of *Sinnlichkeit* can be detected only when sensibility is taken up within the *transcendental* perspective of Kant's investigation. In other words, I suggest that the ambiguity that has accompanied the imagination since Aristotle's positioning of φαντασια between αισθησις and νοησις is dispelled once sensibility is taken up in a transcendental perspective.[9] For, it is only in this framework that the imagination's spontaneous activity can legitimately be ascribed to sensibility without requiring the proviso that distancing it from the passivity and receptivity of the sensible assimilates it to the understanding. Moreover, in arguing for this claim, I also suggest that the active and indeed "productive" role of the imagination as a function of sensibility—a role that famously emerges in §§24–26 of the B edition of the Transcendental Deduction—must lead us to fundamentally re-think the separation that Kant draws between sensibility and understanding in the first *Critique* (and with it the relation between the Transcendental Aesthetic and the Analytic). Ultimately, the very nature of discursive thinking is at stake in this discussion.

It is on this basis that I address the second issue, namely, the connection between the imagination and the task of mediating between sensibility and understanding. That imagination is a form of sensibility excludes both that mediation is the function carried out by a distinct third term placed between *Sinnlichkeit* and *Verstand*,[10] and that imagination is their unknown root as Heidegger suggests. My claim is that the ambivalence that in Kant's presentation seems to always accompany the imagination rests, in fact, on the deeper difficulty that in-

---

9 Sallis (2000, 31) sees in this early Greek positioning of the imagination the root of the problem that this mental "force" carries with itself in its successive history.

10 This does not exclude that the representation produced by the imagination, namely, the schema, can instead be a "third" between the category and the empirical intuition of an object (KrV B177/A138).

habits the concept of mediation in Kant's reflection up to the *Critique of Judgment*. In other words, it is not the case that the imagination is charged with a mediating function because it is constitutively ambiguous in its intermediate position between the sensible and the intellectual. For the opposite is rather the case: transcendentally (or in its transcendental validity), the imagination is unambiguously a form of sensibility, and yet it appears as a fundamentally ambiguous function or activity when it is tied to the task of mediation—and *this* is the problematic concept given the premises of the first *Critique*.

1 *Sinnlichkeit* in Kant's Transcendental Investigation

I begin by framing the novelty of Kant's transcendental view of sensibility in contrast to the modern empiricist and rationalist tradition. This allows me to bring to light the peculiar nature of what is generally referred to as the "dualism" separating sensibility and understanding, concept and intuition that emerges in the first *Critique*, namely, the fact that such dualism rests, in turn, on a deeper distinction that takes place in sensibility itself. In a previous work I have used the interpretive notion of "transcendental embodiment" to indicate the perspective on human sensibility with which Kant overcomes early modern dualisms of mind/soul and body and discloses an unprecedented formal and active aspect of sensibility thereby shifting the dualism within sensibility itself.[11] I shall briefly summarize my argument here in order to present the framework that allows Kant both to count imagination as one of the forms or active functions of *Sinnlichkeit* and to distinguish the reproductive or merely empirical imagination from the productive or transcendental imagination. I argue that the novelty of Kant's position does not consist in adding a new, heretofore unthematized form of the imagination to the traditional repertoire of mental faculties proposed by empirical psychology, but in thinking of sensibility in a *transcendental* perspective, thereby creating the space for a function of the imagination which is itself required if sensibility is to fulfill its peculiar role within the cognitive process, that is, the role of providing a priori conditions for knowledge.

Setting the theory of the first *Critique* against the background of early modern epistemologies and psychology, I maintain that there are two fundamental insights on which Kant's transcendental philosophy ultimately rests. Both are crucial for understanding the distinctive but also problematic character of

---

11 See Nuzzo (2008).

Kant's conception of the imagination. First comes the famous claim, which Kant advances at the end of the introduction, that human knowledge is divided into "two branches," namely, *Sinnlichkeit* and *Verstand*, their difference being a difference in the kind of representations they generate (i.e., intuitions and concepts) as well as in the function that their respective representations play within the cognitive process. Through the former, Kant explains, "objects are *given* to us, while through the latter they are *thought*."[12] The heterogeneity that divides the two branches of human cognition implies that no gradual transition between sensible and intellectual representations is possible, i.e., that no process of clarification can lead from the alleged obscure and confused representations of the senses to the supposedly clear and distinct representations of the intellect. In other words, their difference is "transcendental," not "logical," i.e., does not concern the "logical form" (clear vs. confused) of the representation, but its very "origin and content."[13] Moreover, such heterogeneity also implies that the activity proper, respectively, to sensibility and understanding as mental faculties or as sources of representations is in principle different. Sensibility cannot possibly *think* of objects; the understanding cannot *give* objects to itself. This characterization, while distancing Kant from his early modern predecessors, circumscribes the conditions of *human* knowledge and hints at the situation of mutual dependency in which the two cognitive branches are necessarily set.[14] Thereby it also introduces the central problem of the first *Critique*, namely, the problem of synthesis, and outlines the general partition of the work. Whether these two main constitutive functions of human knowledge originally separate out of the same common "root" is ultimately immaterial since that root remains necessarily "unknown" to us (going back to such an original point would require the impossible gesture of stepping out of the conditions through which all human knowledge is first made possible).[15]

What is more relevant to Kant's project, although it has received far less attention than the remark on the "common although to us unknown root," is the way in which, by this connection, Kant presents sensibility as the possible topic

---

12 KrV B29/A15; see also the "zwei Grundquellen des Gemüts" at B74f./A50f.—this latter passage repeats the former in the distinction between "gegeben" and "gedacht."

13 See KrV B61f./A44 For Kant's rejection and criticism of the Leibnizian-Wolffian version of the early modern gradualistic model, see B61/A44.

14 See KrV B74/A50: "Erkenntnis" requires the contribution of *both* "elements"; and neither is to be privileged over the other (B75/A50).

15 KrV B29/A15.

of a *transcendental* investigation.[16] This is the suggestion from which the critical project takes its departure. Thereby, we get to the second distinctive feature of Kant's theory. Kant announces that a *"transzendentale Sinnenlehre,"* which he will entitle "transcendental aesthetic," can be offered if and only if sensibility is found to be so constituted as to contain "a priori representations."[17] In the *Prolegomena*, Kant insists on the novelty of his position by recognizing that "it never occurred to anyone that the senses might intuit a priori."[18] The idea that sensibility displays an a priori dimension is here set both against a merely empirical view of sensibility and against the *"schwärmerische[r] Idealismus"* of intellectual intuition, which ultimately assimilates the sensible to the intellectual.[19] In putting forward this proposal, Kant drives a deeper separation—another, more fundamental dualism, as it were—within sensibility itself. It is on this hypothesis, from which the thesis of transcendental idealism depends, that the project of the first *Critique* ultimately rests. Now, I have suggested that it is precisely the distinction between a material and empirical, and a formal hence a priori and transcendental aspect of sensibility that allows Kant to overcome the modern mind-body dualism, but also fundamentally complicates its aftermath.[20] A first complication can be immediately detected just by looking back to that famous introductory claim on the distinction of the two cognitive branches. If sensibility does entail an a priori aspect (or is capable of producing a priori representations of its own), then the heterogeneity that divides the sensible and the intellectual cannot be construed by assigning a merely a posteriori character to the former while leaving apriority as the exclusive province of the latter. Since both *Sinnlichkeit* and *Verstand* turn out to host and be the source of a priori representations, Kant needs a stronger ground for separating the two than what was available to his predecessors. This problem, I suggest, constitutes the basis for Kant's new distinction between intuition and sensation (and between pure and empirical intuition), but is also the basis for his introduction of a productive imagination next to the merely reproductive one.

Although Kant seems comfortable with taking up the traditional language that presents sensibility as "receptivity" and as a fundamentally passive faculty

---

16 In this passage the possibility that the understanding can indeed undergo transcendental investigation is not even raised. The fact that Kant concentrates exclusively on sensibility betrays his awareness of the novelty of the project precisely in this respect.

17 KrV B30/A16.

18 *Prolegomena*, A 207, AA IV, 375 Fn.

19 *Prolegomena*, A 207, AA IV, 375 Fn.

20 See Nuzzo (2008) Part I for a discussion of Kant's relation to the early modern mind-body dualism.

(as a capacity for being "affected"), thereby easily contrasting it to the understanding's activity or "spontaneity," he neither reduces *Sinnlichkeit* to the mere receptivity of the senses nor does he view it as merely inert passivity.[21] Here again, while this position requires Kant to provide a stronger ground for the separation between sensibility and understanding than simply casting the former as material and the latter as formal, it offers a persuasive basis for distinguishing two fundamentally different aspects of sensibility itself, namely, one which is material, passive, and receptive, the other which is formal, active, and somehow even endowed with a spontaneity of its own. The point, however, is that the latter can be detected and thematized only on the condition of endorsing a transcendental perspective. Indeed, in light of these considerations, we can recognize a crucial ambiguity in Kant's initial claim that sensibility is the power "through which objects are given to us."[22] While the empiricist could easily subscribe to this claim by underlining the passive function of the senses through which given material objects affect us, Kant somehow reverses the relation: sensibility, if transcendentally considered, displays an aspect whereby it is the *activity* producing representations that make it possible for objects to be given to us in the first place (as objects of possible experience). For, such representations provide the a priori formal conditions under which alone objects can be represented *as given*. In the first case the givenness of the object is the condition for its empirical representation; in the latter the a priori representation is the condition for the givenness of the object to our mental faculty. Thus, whereas for the empiricist the claim simply confirms the aposteriority and object-dependence of the representations of the senses, for Kant it expresses the extent to which sensibility, by displaying an a priori aspect, is the topic of a transcendental investigation, and confirms the 'sensibility-dependence' of appearances as objects of possible experience. Accordingly, at the beginning of the Aesthetic, Kant distinguishes the formality and apriority of pure intuition (*reine Anschauung*) from the materiality and aposteriority of both sensation (*Empfindung*) and empirical intuition (*empirische Anschauung*).[23] On this basis he brings to light the specific cognitive function of space and time as a priori forms of pure sensible intuition. Transcendentally, the forms of intuition or the formal aspect of sensibility *precedes* the givenness of objects and makes it possible.

---

21 KrV B33/A19, also B74/A50, which distinguishes "Rezeptivität der Eindrücke" and "Spontaneität der Begriffe," and B75/51.

22 KrV B29/A15 and B33/A19.

23 KrV B34 ff./A19 ff.; see again at the beginning of the Analytic B74/A50: in this case "pure" and "empirical" qualify both intuitions and concepts.

It is striking, in this regard, that a feature of the imagination, which since Aristotle is specifically used to define this mental activity, once shifted or transformed within the transcendental perspective (or *"im transzendentalen Verstande"*)[24] serves Kant to characterize the formal and pure aspect of sensibility in contrast to its material and empirical side. If the imagination is traditionally defined as the capacity to represent objects without their being immediately present to the senses, the same relation now appears as a specification of the general a priori aspect of sensibility, namely, its preceding the givenness of objects (and, this time, its making such givenness possible and cognitively meaningful). What obviously changes with Kant's transcendental turn is the meaning of the independence that the formal aspect of sensibility (and the imagination, transcendentally considered) enjoys with regard to the object of sense and empirical affection. What is relevant, however, is the closeness that from the outset the transcendental investigation establishes between imagination and the (pure) form of intuition. Kant introduces the notion of a pure form of sensibility, namely, the condition on which hinges the very possibility of a "transcendental aesthetic" (or of a *"transzendentale Sinnenlehre"*), by inviting us to make abstraction, in the representation of a body, from all that belongs to our thinking as well as from all that belongs to our sensation and empirical intuition of a body. Kant concludes that what "still remains over from this empirical intuition, namely, extension and figure" belongs to "pure intuition." Now he maintains that *"even without any actual object of the senses or of sensation"* pure intuition "exists in the *Gemüt* a priori as a mere form of sensibility."[25] In other words, pure intuition is the a priori form or formal condition that, independent of the actual presence of the object, first allows objects to be perceived and represented as present. Pure intuition is the form within which all sensation is inscribed. Significantly, a long-standing tradition characterizes the imagination in similar terms (thereby bringing imagination close to memory): φαντασια, maintains Aristotle, is the capacity to preserve what is absent and recall its presence; it is ειδος without υλη.[26] Kant himself echoes the traditional definition, which repeatedly appears in the handbooks of empirical psychology of the time,[27] when he introdu-

---

24 KrV B34/A20.

25 KrV B35/A21 (my emphasis).

26 Again, to appreciate such similarity, we should make abstraction from Kant's transcendental turn. See for example Aristotle, *De anima*, 432a 9–10; *De memoria et reminiscentia*, 449b 24–450a 6; see Ferraris (1996, 37–40); Ferrarin (1995b); Frede (1992); Mörchen (1930).

27 See Ferrarin (1995b, 69f.) who refers to Aristotle, *De memoria et reminiscentia*, 449b 24–450a 6 but immediately proceeds to discuss the historical differences separating Kant and Aristotle.

ces the imagination as "the faculty of representing in intuition an object that is not itself present."[28] While the transcendental characterization of the pure form of sensibility as that which in sensibility precedes the giveness of the object is a distinctively Kantian discovery, the empirical characterization of the imagination as the capacity to represent an object that is not (or is no longer) present to the senses is the way in which an established tradition defines the imagination through its intermediate position between the material world of objects and the independent activity of the mind. On the basis of this definition, given the pre-Kantian homogeneity that underlies sensible and intellectual representations distinguishing them only through degrees of clarity and distinctness, the imagination is alternatively construed as a form of thinking (i.e., relatively independent of given objects, yet necessarily referred to material objects as a "special way of thinking of material things," as Descartes puts it)[29] or as a form of sensibility (less close to the materiality of the object, merely "decaying sense," as Hobbes puts it).[30] By contrast, in the framework of Kant's transcendental investigation, the traditional definition of the imagination conveys a new meaning as the imagination now falls squarely and unambiguously within the pure, a priori and formal dimension of *Sinnlichkeit*. For, if sensibility is indeed the faculty through which objects are given to us, to the extent that such faculty displays an a priori dimension, its forms or principles are independent of and prior to the givenness of the object. Accordingly, independence from the actual presence or givenness of the object is no longer a sign of the imagination's ambiguous position between thinking and the senses. Rather, it indicates, transcendentally, the imagination's belonging to the *formal* aspect of sensibility itself. Imagination is now brought close to intuition—and to the bifurcation that divides empirical and pure intuition. It is precisely this new proximity that requires Kant to transform the empirical notion of imagination, and to introduce a parallel bifurcation within imagination itself. Ultimately, the root of Kant's distinction between a merely empirical and reproductive imagination and a transcendental or productive imagination does not lie primarily in the opposition between *Sinnlichkeit* and *Verstand*, but, rather, in the deeper dualism which Kant recognizes within the province of *Sinnlichkeit*, namely, in the separation of a formal and a priori, and a material and a posteriori function of sensibility. In this way, the distinction that the transcendental inquiry draws between *Empfindung*, which is always material, and *Anschauung* which allows for pure a priori forms (in the distinction

---

28 KrV B151—I shall analyze this definition in more detail below.
29 See Descartes, *Meditation* VI, AT 6, 36; see Lyons, John D. (1999); Sepper (1989).
30 Hobbes, *Leviathan*, I, 2: "Of Imagination"; see Sepper (1988).

between empirical and pure intuition) is paralleled by the distinction that separates the reproductive and merely associative imagination and the properly transcendental and productive imagination. While the former is the topic of empirical psychology, the latter can be addressed only by transcendental philosophy.[31] In order to get to Kant's characterization of the *productive* imagination, however, a further step in the argument is needed.

2 Transcendental Aesthetic, Imagination, and the Thesis of Transcendental Idealism

The discovery of pure a priori forms of sensible intuition allows Kant to develop a specific "science of all the a priori principles of sensibility." This science, under the title of Transcendental Aesthetic, constitutes the first division of the *Critique of Pure Reason*. By rejecting the contemporary German use of the term "*Ästhetik*," which refers to what is at the time called a "critique of taste," hence refusing to follow Baumgarten in the attempt to develop a critique of taste in scientific form (for taste, being merely empirical, cannot be brought back to a priori principles), Kant harkens back instead to the ancient Greeks who divide knowledge in αισθητα and νοητα (the sensible and the intelligible).[32] Methodologically, the "transcendental aesthetic" as a transcendental theory of all the a priori forms of sensibility proceeds by "isolating" sensibility from the activity of the understanding and its concepts so that only "empirical intuition" remains. But it also proceeds by isolating, within sensibility, "pure intuition" from all sensation so that only the formal elements of sensibility and only the "form of appearance" which they yield are left.[33] As such an "isolated" discipline, the Transcendental Aesthetic famously exhausts the realm of pure sensibility by presenting space and time as the only pure a priori forms of sensible intuition. No mention of the imagination is made in the Aesthetic. However, it should be underscored that the imagination is not excluded from the Aesthetic on the ground that it

---

31 See KrV B152. This overall position, which connects imagination to intuition as the a priori formal side of sensibility, thereby complementing the empiricist and psychological treatment of sensibility with a transcendental account, is confirmed among other texts by the *Vorlesungen über Metaphysik* (*Metaphysik v. Schön*) in which Kant claims: "To empirical intuition belongs sense (*Sinn*); to pure intuition *imagination* (*Einbildungskraft*). The latter is the capacity for intuition even in the absence of objects. Both together, sense and imagination, constitute sensibility" (AA, XXVIII/1, 472f.).

32 See KrV B35f./A21 Fn.

33 KrV B36/A22.

does not belong to sensibility. In fact, as suggested above, a covert reference to the imagination is made through the means by which, in introducing his transcendental aesthetic, Kant circumscribes the a priori dimension of sensibility as such. Rather, the reason why the imagination is not thematized in the transcendental aesthetic lies in the procedure of 'isolation' on which its analysis rests. Indeed, the theory of sensibility of the first *Critique* is not restricted to the Transcendental Aesthetic, but receives a fundamental extension in the Analytic. For, the imagination is the faculty that brings pure intuition out of its 'isolation' by connecting it to the activity of the understanding and to the actual givenness of objects. Thereby, in bringing to the fore the activity of the imagination Kant revisits the Aesthetic this time no longer taking its forms in 'isolation', but this time in connection with the judging function of the understanding. Thus, if the Transcendental Aesthetic does not make reference to the imagination, when Kant introduces this faculty in the Analytic the argument of the Aesthetic is taken up and expanded with regard to the role that sensibility plays in the broader cognitive process. However, before getting to this further step of the argument, we need to mention yet another respect in which the imagination as a constitutive component of sensibility is present, albeit indirectly, in the Transcendental Aesthetic.

The Transcendental Aesthetic establishes the claim on which the entire edifice of the first *Critique* (and furthermore its relation to the second *Critique*) depends, and which sets Kant apart from every one of his empiricist and rationalist predecessors. This is the thesis of transcendental idealism.[34] According to this thesis, space and time are subjective, a priori forms of our sensible intuition (they are, respectively, the a priori form of "outer" and "inner sense"). As such they are transcendentally ideal and empirically real. They are neither intellectual representations belonging to the understanding (i.e., they are not concepts)[35] nor properties of things in themselves. On the former condition depends the separation between the sensible and the intelligible, hence the exclusion of the possibility of an intuitive understanding, which would immediately give objects to it by simply thinking of them. The latter condition implies that objects that have no relation to the form of our sensibility (for example, things in themselves) are neither spatial nor temporal. As a priori forms of intuition space and

---

34 It is not my present aim to provide a detailed discussion of the thesis of transcendental idealism but only to connect such thesis with the issue of the absence of the imagination in the Transcendental Aesthetic. For the novelty of transcendental idealism with regard to the tradition, see the already mentioned passage of *Prolegomena*, A 207, AA IV, 375 Fn.

35 See the third argument on space and the fourth argument on time (respectively, KrV B39/A24f. and B47/A31f.)

time are the formal a priori conditions for all appearance in general. To this extent they are *transcendentally ideal*—i.e., if abstraction is made from the subjective conditions under which objects are given to us, time and space are "nothing";[36] and they are *empirically real*—i.e., they have "objective reality" or "objective validity" with regard to "all objects that allow of ever being given to our senses" (time) or with regard to "all that can be presented to us externally as object" (space).[37]

In their transcendental ideality and empirical reality, space and time are the conditions of the "receptivity of our *Gemüt*." They "contain (*enthalten*) a manifold of pure a priori sensibility," thereby providing the "material" (*Stoff*) to the concepts exposed in the transcendental logic. In this way, the Aesthetic furnishes the distinctive "content" (*Inhalt*) that sets Kant's *transcendental* logic apart from traditional general logic (which, instead, makes abstraction from all content as such).[38] Properly, however, space and time are only the pure forms "under which something is intuited,"[39] i.e., the pure forms under which all empirical intuition and sensation takes place and in which the sensible manifold is given (they "contain" it). While this implies that these forms "exist in the *Gemüt* a priori" independently of what is being sensed or intuited,[40] as remarked in the Aesthetic, it also implies that in their formality they do not directly furnish the material to the understanding's concepts—at least not in the 'isolated' way in which they are presented in the Aesthetic. It is at this point that the imagination is called into the picture. The imagination explains how space and time can provide the material for the understanding's concepts—how they can contain a manifold of intuition.

It is first within the forms of pure intuition that the sensible manifold is received by the *Gemüt* as structured by relations of spatial juxtaposition and temporal succession. Yet, in order to be taken up and represented in this way (i.e., as spatial and temporal), the manifold of empirical intuition needs to be brought together by the activity of synthesis. It is here (not in the pure forms of intuition as such, i.e., in these forms taken in isolation) that we find the "first origin (*den ersten Ursprung*) of our knowledge." Now, Kant maintains that "synthesis in general (*Synthesis überhaupt*) [...] is the mere action/effect of the imagination (*die bloße Wirkung der Einbildungskraft*)."[41] That the imagination, precisely in its pro-

---

36 KrV B52/A36 for time; B 44/A28 for space.
37 KrV B52/A35 f. for time; B 44/A28 for space.
38 KrV B102/A76 f.
39 KrV B74 f./A50.
40 KrV B35/A21 (my emphasis).
41 KrV B103/A78.

ducing "synthesis in general" belongs to sensibility is confirmed in this same passage by the qualification of "blind" with which Kant contrasts the imagination to the understanding—*Sinnlichkeit* to *Verstand*.[42] The task or "function" of the understanding is to "bring this synthesis [i.e., the synthesis of the imagination] to concepts."[43] Once the Aesthetic is concluded and the procedure of 'isolation', having brought to the fore space and time as its two a priori forms, has exhausted its function, it becomes clear that what is spatial and temporal (and on the way to become the content of the understanding's concepts) is not sensation itself but the imaginatively synthesized manifold of sensation. But if it is so, then an important question arises at this point (namely, at the beginning of the Analytic of Concepts). Granted that the thesis of transcendental idealism claims that all things which do not fall under the forms of pure intuition—hence, undoubtedly, things in themselves—are a-temporal and a-spatial, are sensations in themselves, i.e., before being taken up under the pure forms of intuition, also a-temporal and a-spatial?[44] It seems that outside and before—and in isolation from—the imaginative synthesis space and time, being only the forms "under which something is intuited,"[45] belong neither to things in themselves nor to sensations in themselves. Indeed, since it is the imagination that brings the manifold of sensations under a pure intuition (thereby making sensations representable in the order of succession and juxtaposition), the thesis of transcendental idealism should be seen as a thesis that regards the pure dimension of *Sinnlichkeit* in its entirety, namely, space, time, *and the imagination*—to the extent, that is, that the latter produces "pure synthesis" in general, i.e., that it makes the pure form of intuition into the "manifold of a priori sensibility" which is the possible content of the category.[46] In this case, the transcendental ideality and empirical reality implied by transcendental idealism regard not just

---

42 KrV B103/A78; recall B75/A51. "Blindness" qualifies intuitions and the imagination to the extent that they are set in relation (or in the absence of relation) to concepts. It is not, therefore, a designation that we can expect to find in the Aesthetic where intuition is not set in relation to concepts.

43 KrV B103/A78.

44 Presently, I am interested in this question, which has extensive ramifications in the Analytic, only insofar as it is relevant to our present topic. In this argument, I follow, in part, Waxman (1993, 67–69), who develops his interpretation of Kant's transcendental idealism with regard to the Analogies of Experience.

45 KrV B74f./A50.

46 This claim is confirmed by the first of the three-fold syntheses of the A deduction, see KrV A99: at stake is the unification of the manifold that the "sensibility offers in its original receptivity." Such unification-synthesis is what allows the representations of space and time to be given a priori.

space and time but space and time in relation to the blind synthesis of the imagination: it is only to the extent that our perceptions of things are synthesized by the imagination under the pure forms of intuition that things appear to us in the order of succession and juxtaposition. This means that not only things in themselves but also sensations that are not imaginatively synthesized under the pure forms of intuition are a-spatial and a-temporal; or, that if abstraction is made from the way in which the imagination brings the manifold of sensation under the form of intuition, space and time are nothing for us: prior to and independently of imagination, sensations are transcendentally real.[47] Indeed, if the thesis of transcendental idealism is a thesis that concerns the a priori, formal aspect of *Sinnlichkeit* as condition for the givenness of objects, and if, as I have claimed above, the imagination (insofar as it yields "synthesis in general") is one of the functions of a priori sensibility (next and connected to pure intuition), then it should not be surprising that the validity of transcendental idealism extends to space, time, *and the imagination.*

3 The Transcendental Aesthetic Revisited: Synthesis Speciosa and the Productive Imagination

The imagination is first introduced in the presentation of the activity that gathering the manifold of intuition under the pure forms of space and time brings the latter out of the 'isolation' required by the analysis of the Aesthetic, and makes them into the possible content of the understanding's concepts. Now, by presenting the imagination as the activity of "synthesis in general" Kant brings to light the fundamentally *active* aspect of sensibility—an aspect that is not yet directly tackled in the Aesthetic where the a priori character and formality of intuition are mostly at issue. Synthesis is *"Handlung."*[48] Significantly, the "receptivity" of the *Gemüt* and the "action" of the imagination that produces synthesis as its "effect," are now perfectly compatible qualifications of sensibility transcen-

---

47 See Waxman (1993, 68): "transcendental idealism involves the denial nor merely of supersensory reality to space and time but superimaginative reality as well." The problem raised by this interpretation of transcendental idealism is outlined by Waxman (1993, 69). An extensive discussion of the literature is in Banham (2006) chapter 1.
48 KrV B102/A77.

dentally considered.[49] This implies, yet again, that the distinction between sensibility and understanding cannot be construed as the mere opposition of passivity and activity, and is not exhausted by the opposition of receptivity and spontaneity but requires the more complex distinction of different types of activity—and even of different types of synthesis.[50] And this leads us to the crucial intervention of the imagination in the transcendental deduction.[51] It is at this point that the imagination's activity—and properly even its transcendental productivity and spontaneity—gains the center stage in explaining how the understanding's concepts can refer or be applied to "objects of the senses in general,"[52] thereby triggering a fundamental re-visitation of the Transcendental Aesthetic.

In the B edition of the Transcendental Deduction, the argument of the Transcendental Aesthetic on the nature of space and time is successively revisited in light of the principle of the synthetic unity of apperception and once the activity of the imagination and its relation to the understanding come to the forefront.[53] In §17 the principle of the Transcendental Aesthetic—that the manifold of intuition should "be subject to the formal conditions of space and time"—is first expanded in relation to the principle of "all use of the understanding" to state that "all manifold of intuition should be subject to the condition of the original synthetic unity of apperception."[54] While the former principle is the condition of the *givenness* (*gegeben*) of the represented manifold, the latter is the condition of its *combination* (*Verbindung*) in one consciousness.[55] With this claim, Kant restates the division of labor between sensibility and understanding within the cognitive process and also offers an insight into the presence within sensibility of the activity of synthesis. Space and time are now presented not simply as forms of intuition, but as intuitions or "singular representations" that contain the unity of a

---

49 KrV B103/A78. Indeed, it is the imagination's action that reveals the cognitive significance of the *Gemüt*'s receptivity.

50 Ferrarin (1995a, 142f.) concludes that the imagination as power of "synthesis" hence as activity belongs to sensibility *as well as* to the understanding. By contrast, I argue that it belongs to sensibility, which transcendentally considered involves an activity. In other words, activity as such is not the exclusive province of the understanding. See also Banham (2006, 11 ff.).

51 My aim here is neither to discuss this enormous issue on which the literature is as vast as on no other nor to tackle the problem of the shift in the role that the imagination plays in the deduction from the A to the B edition. My argument remains confined to the claim that the imagination belongs to sensibility. I shall concentrate on the crucial moments of the B deduction in which Kant's re-visitation of the Transcendental Aesthetic is most clearly thought out.

52 KrV §24 title.

53 KrV §17 and §§24 – 26.

54 KrV§17 B136.

55 KrV§17 B137.

manifold and imply the unity of the consciousness of a manifold of representations.[56]

In §24 and then in §26, dealing with the "application of the categories to objects of the senses in general,"[57] or to the sensible given, Kant returns to a discussion of the forms of intuition. In order to do so, he appeals to the imagination granting it a crucial role in bringing the deduction to conclusion, and in offering, at this point, a new take on the argument of the Transcendental Aesthetic.[58] At stake is the transition from the purely intellectual synthesis contained in the categories, as "mere forms of thought, through which alone no determinate object is known," to the application of the categories, "to objects that can be given to us in intuition," i.e., to appearances. In the latter case alone will it be proved that the understanding's concepts have "objective reality."[59] To this aim, Kant distinguishes two forms of transcendental synthesis. He distinguishes the "figurative synthesis" or *synthesis speciosa*, i.e., the "synthesis of the manifold of the sensible intuition" from the "*Verstandesverbindung*," i.e., the intellectual combination or *synthesis intellectualis* that is "thought in the mere category in respect to the manifold of an intuition in general."[60] Both the category and the sensible intuition *entail* a synthesis—the former is intellectual and still indeterminate or empty (still no knowledge of a *determinate* object, referred as it is to the manifold of an intuition *in general*); the latter is figurative, referred to the manifold of the *sensible* intuition and still in need to be thought. Moreover, it is clear that both sensibility and understanding *display* the activity of synthesis; but also, and more importantly, that both faculties *produce* synthesis. For, the *synthesis speciosa* describes the activity of the imagination in its productive, i.e., transcendental function.[61] In point of fact, Kant distinguishes the "figurative synthesis" from the "intellectual combination" by assigning the former to the imagination as a sensible faculty. The "figurative synthesis, if directed merely to the original synthetic unity of apperception, that is, to the transcendental unity which is

---

56 KrV§17 B136 Fn.—this refers to the claim of the Aesthetic that space and time are intuitions and not concepts.

57 KrV§24 title.

58 See Longuenesse (2000, 211–242, 213): "The goal of the Transcendental Deduction of the categories 'is fully attained' only when it leads to a rereading of the Transcendental Aesthetic."

59 KrV§24 B150 f.

60 KrV§24 B151. See Longuenesse (2000, 211 ff.) who underscores the continuity of Kant's position from the *Dissertatio* to the *Critique*. In the former Kant argues that space and time are "*formae seu species*" that belong to sensibility (AA II, 392f; 384 f.); in the passage we are now analyzing *synthesis speciosa* belongs to the imagination. The continuity consists in Kant's using the latter to further his theory of sensibility. See also Makkreel (1990, 26–42).

61 KrV B152.

thought in the categories, must [...] be called the *transcendental synthesis of the imagination.*"[62] In §10 the presentation of the categories is connected to the "function" of the understanding, which "brings to concepts" the blind imaginative synthesis of the manifold contained in the pure forms of intuition. At this juncture, instead, when the task is to explain the "application" of the categories to objects given in intuition, the synthesis of the imagination refers to the transcendental unity thought in the category and provides for it the *determinate* sensible intuition which the concept is still lacking (for, through it "no determinate object is known").[63] In §10 "synthesis in general" is defined as the *"Wirkung der Einbildungskraft,"* the "blind" power belonging to sensibility and indispensable for cognition. In §24, as the *synthesis speciosa* is distinguished from the *synthesis intellectualis* on the ground of its being the "transcendental synthesis *of the imagination,"* Kant seemingly appeals to the traditional definition of this mental power, yet also fundamentally modifies it. "Imagination is the faculty of representing in intuition an object that is *not itself present.*"[64] Unlike the tradition, which stresses the imagination's capacity to represent an object even when it is not present to the senses, Kant directly connects the imagination to intuition. The absence of the object, on the other hand, is not an empirical but a transcendental absence (transcendentally, the intuition precedes the object and makes its givenness possible). The imagination's task is to *"represent in intuition"* an object not immediately present, i.e., to furnish the intuition through which the sensible object is given to which the category applies.[65] By providing the determinate intuition whereby the concept applies to the sensible given object—and by providing the intuition even though the object is not present—the imagination is indispensable to understand the possibility of the concept's *application* to objects. It is at this juncture that it becomes crucial to stress the sensible nature of the imagination (or that imagination is a power belonging to sensibility)—and to hold on to this thought despite the ambiguity of Kant's text. For, if the imagination is viewed as a function of the understanding (or indeed as identical with the understanding) the claim amounts to assigning to the intellectual faculty a ca-

---

62 KrV B151.

63 KrV B150. In other words, the relation here is the reverse of the relation established in the presentation of the categories: there the understanding "brings to concepts" the blind synthesis of the imagination, i.e., the manifold of intuition contained in the pure forms of intuition (B103/A78); here the imagination's figurative synthesis concerns the transcendental unity thought in the category (B151).

64 KrV B151.

65 Indeed, this definition makes sense only on the basis of Kant's transcendental theory of sensibility (and of intuition in particular).

pacity for intuition, i.e., the capacity of giving itself objects. And from this possibility, namely, from the possibility of an intuitive understanding Kant here (as in all places where such a possibility seems to even remotely surface) immediately distances himself. To this effect he maintains that "since all our intuition is sensible, the imagination, owing to the subjective condition under which alone it can give to the concepts of the understanding a corresponding intuition, belongs to *sensibility*."[66] The imagination provides the understanding, which is not itself intuitive, with a corresponding sensible intuition as it represents *in intuition* the (absent) object. This is the condition for the "application" of the category to the sensible given.

The figurative synthesis of the imagination is not just this faculty's "*Handlung*." It is the "exercise of spontaneity (*Ausübung der Spontaineität*)," and as such it is not merely "determinable (*bestimmbar*)," as are receptivity and the material part of sensibility, i.e., "sense." Again, the imagination in its capacity to spontaneously produce synthesis belongs to the pure, active aspect of *Sinnlichkeit*. The imagination is not merely reproductive and associative in its activity but genuinely productive. The spontaneity exercised in the synthesis of the imagination is "determining (*bestimmend*)." Indeed, the imagination is able to determine "sense" (the inner sense) a priori "in respect of its form in accordance with the unity of apperception."[67] To this extent, the imagination is presented as the "faculty which determines the sensibility a priori."[68] In other words, there is a determining spontaneous aspect of sensibility, i.e., the productive imagination, whereby sensibility, at once receptive (sense) and spontaneous (imagination), *determines itself*—this is the idea of *Selbstaffektion* that Kant addresses, albeit obliquely, already in the Transcendental Aesthetic.[69] How do the two presentations of the imagination offered in this passage—namely, its capacity to represent in intuition the non-present object thereby providing the category with a corresponding intuition, and its capacity to determine sensibility a priori—relate to each other in the idea of the figurative synthesis? While the *synthesis speciosa* is distinguished from the *synthesis intellectualis* by means of the faculty that produces it, the very nature of such synthesis connects the imagination to the unity of apperception. The "synthesis of intuitions *in accordance with the categories*"— argues Kant—"must be the transcendental synthesis of the *imagination*." This transcendental synthesis is a synthesis of sensible intuitions that being "directed

---

66 KrV B151.

67 KrV B151f.

68 KrV B151; at B161 Fn. Kant refers to the understanding as determining the sensibility through the synthesis of the imagination (see the discussion below).

69 See KrV B152 which refers back to §6.

to the original synthetic unity of apperception" and determining "sense" formally and a priori "in accordance with the unity of apperception" is also, by consequence, in accordance with the categories. It is at this point, where the imagination's synthesis connects to the understanding's concepts that many interpreters see the imagination's allegiance to sensibility waver. [70] The synthesis of the imagination, Kant announces, is "*eine Wirkung des Verstandes auf die Sinnlichkeit*"—is the action/effect of the understanding on sensibility—"and is its first application [...] to objects of our possible intuition."[71] The identity of these two actions is significant here: the understanding's *effect* or *efficacy* on sensibility *is* its first *application* to objects of intuition. But why is the figurative synthesis, which heretofore has been repeatedly assigned *to the imagination*, now being presented as the "*effect of the understanding* on sensibility"? The imagination is the sensible source or power producing the figurative synthesis—Kant never revokes this claim. The product of this imaginative activity, however, reveals the *Wirkung* of the understanding because the imagination's production of synthesis is functional to the understanding's activity of judgment. Since the synthesis of the imagination provides the category with a corresponding intuition, it offers the first application (and the basis of all application) of the concept to appearances. At this point, however, Kant looks at the process from the side of the understanding and its own activity. To the extent that it is set to determine intuition for the category, sensibility's imaginative self-determination is properly the "effect of the understanding." And to the extent that the synthesis of the imagination is "in accordance with the categories," it displays the understanding's determination while still belonging to the imagination (and being the result of the imaginative spontaneity). The presence of the imagination and its spontaneous activity is, Kant reiterates, precisely what distinguishes the "figurative" from the "intellectual" synthesis in which the imagination plays no role. It is neither the ground for a straightforward assimilation of the imagination's activity (including its spontaneity) to the function of the understanding, nor the ground for assigning to the imagination an intermediary role between *Sinnlichkeit* and

---

70 KrV B152. See for example Long (1998, 237–240, 248 f.), who fundamentally misunderstands this passage: it is not the imagination as a faculty but its synthesis to which Kant here refers as "Wirkung des Verstandes auf die Sinnlichkeit."

71 KrV B152. See for example Long (1998, 237–240, 248 f.), who fundamentally misunderstands this passage: it is not the imagination as a faculty but its synthesis to which Kant here refers as "Wirkung des Verstandes auf die Sinnlichkeit."

Verstand.[72] It expresses instead Kant's most advanced presentation of the formal and spontaneous aspect of sensibility, an aspect that no psychological analysis but only "transcendental philosophy" succeeds in bringing to light.[73]

Indeed, the notion that sensibility is both determining and active (through the imagination), and determined and such as revealing in its synthesis the "effect" or action of the understanding seems a "paradox." It is, instead, only an implication of the thesis of transcendental idealism with regard to the inner sense—i.e., the fact that the inner sense represents to consciousness only "the way in which we appear to ourselves, not the way in which we are in ourselves, because we intuit ourselves only as we are inwardly affected"[74] which seems to involve, at the same time, an activity and a state of passivity. And it is an implication of the radical separation—yet also of the mutual interdependence—of sensibility and understanding, i.e., once again, a consequence of the fact that our understanding is not intuitive but needs the imagination to "represent in intuition" its objects.[75] The understanding and the original unity of apperception determine the inner sense through the figurative synthesis of the imagination;[76] the figurative synthesis, on the other hand, lends the understanding's concepts a *determinate* intuition. At stake is the difference between the simple thought "*that* I am"—a representation that is "a thought, not an intuition," and implies intellectual synthesis—and the representation of "a determination of my existence" through "a determinate type of intuition," which requires the figurative synthesis whereby the sensible manifold is given to the inner sense.[77] To comprehend this, however, apperception and inner sense must be recognized as fundamentally distinct, argues Kant against the "systems of psychology"[78] of his time. And it is relevant that what these systems erase in conflating inner sense and apperception is precisely the distinction between the sensible and the intellectual. Hence, only under the condition of keeping them separate can the figurative synthesis be viewed as the bridge or activity that connects the two in producing knowledge (thereby avoiding both the position of psychology which reduces the imagination to its empirical, merely reproductive use, and the position of *Schwärmerei*

---

72 See, for example, the ingenious way in which Long (1998), with the help of Aristotle, construes the ("dynamic") "identity" of imagination and understanding; Long responds to the difficulty already encountered in Allison (1983, 162 ff.); see also Ferrarin (1995a, 142).

73 See KrV B152.

74 KrV B152 f.

75 A point that Kant repeats, yet again, in this connection: KrV 152 f.

76 KrV B153: "Das, was den inneren Sinn bestimmt, ist der Verstand."

77 KrV §25 B157.

78 KrV B153.

which endows the understanding with an intuitive capacity). Echoing the opening of §24, Kant clarifies that the synthetic unity of apperception, as the "source of all combination," refers to the "manifold of *intuition in general*," and in the categories, "prior to all sensible intuition, to objects in general"; the inner sense, by contrast, contains the mere "*form* of intuition, but without combination of the manifold in it," hence does not contain any "*determinate* intuition." How, then, do we move from the manifold of intuition (and objects) "in general" to the "*determinate* intuition" to which the concept is applied? What is required at this point is the "consciousness of the determination of the manifold by the transcendental act (*Handlung*) of the imagination," namely, the figurative synthesis. This synthesis, which Kant previously indicated as the "effect" (*Wirkung*) of the understanding on sensibility, he now qualifies as its "influence (*Einfluß*) on the inner sense."[79] It is the figurative synthesis that *determines* (and gives unity to) the manifold of intuition, a determination that neither the concept as mere *Gedankenform*[80] nor inner sense as mere "*form* of intuition" could display.

The argument with which Kant concludes §24 leads to the crucial issue of the distinction between "form of intuition" and "formal intuition" in §26. Thereby, the B deduction offers a fundamental extension of the theory of space and time of the Transcendental Aesthetic and brings the Transcendental Deduction to its end.[81] Here again, what interests me is solely the extent in which Kant connects the imagination to intuition thereby bringing to light a fundamental aspect of *Sinnlichkeit* transcendentally considered. Space and time, contends Kant, "are represented a priori not merely as *forms* of sensible intuition, but as themselves *intuitions* which contain a manifold, and therefore are represented with the determination of the unity of this manifold (*vide* the Transcendental Aesthetic)."[82] While the Transcendental Aesthetic presented space and time as forms of intuition, now Kant reveals that they are themselves *intuitions* because, in contrast to the 'isolationist' procedure of the Aesthetic, he presently considers space and time in the context of the connection offered by the synthetic activities of imagination and understanding. Kant expresses his complex re-visitation of the Aesthetic in a famous footnote.[83] Here he offers a sort of regressive argument that brings to light the presupposition on which the claim of the Aesthetic that

79 KrV B154.

80 KrV B150.

81 For an extensive reconstruction of the new import of Kant's position here see Longuenesse (2000), chapter 8; see also Baum (1986); Banham (2006) chapter 4; Kitcher (1986); Waxman (1991), chapter 2.

82 KrV §26 B160—a punctual comment on this passage is in Longuenesse (2000, 215 f.).

83 KrV §26 B160 f. Fn. See Kitcher (1986).

space is a "mere form of intuition" rests. Now space itself is represented as "object" (*Gegenstand*) (this is the case, Kant notices, in geometry), which implies more than is warranted by the mere form of intuition. It implies the action of "gathering together"—the *Zusammenfassung*—"of the manifold, given according to the forms of sensibility, in an *intuitive* representation." The Aesthetic only showed how the manifold is *given* according to the forms of intuition; now at stake is the way in which that given manifold—precisely in order to be given— is *gathered together* in a "formal intuition." The latter action comes before the former. While the "*form of intuition* gives only a manifold, the *formal intuition* gives unity of representation." To be sure, the Aesthetic did not just make abstraction from such unity. It did not, however, thematize or explain its transcendental origin, for this origin is the imagination of which no account can be given yet in the Aesthetic. For this reason, in the initial step of his theory of sensibility, Kant assigned such unity "to *Sinnlichkeit*, simply to emphasize that it *precedes* any concept, although as a matter of fact it *presupposes* a synthesis which does not belong to the senses (*Sinnen*)."[84] This much is revealed at this point: the unity present in space is the unity belonging to sensibility not to the understanding; yet this unity belongs to sensibility taken in its formal and active dimension—for unity always implies the activity of unification and this can never be found in the "senses" (i.e., in the material aspect of sensibility). While such sensible unity *precedes* the concept, it rests, in its turn, on a more original "synthesis," which itself makes possible "all concepts of space and time," i.e., space and time as objects (as in geometry).[85] This more original synthesis is the *synthesis speciosa* introduced in §24. Now Kant claims that in it "the understanding determines the sensibility (*der Verstand die Sinnlichkeit bestimmt*)," (insofar as through this synthesis space and time are first *given* as intuitions),[86] yet its unity is not the unity of the understanding's concept, but a unity that belongs to space and time themselves precisely as a priori intuitions.[87] Since here the understanding, as the capacity to form judgments, "determines sensibil-

---

84 KrV B160f. Fn. (my emphasis).

85 See also AA XXII, 76: with regard to space and time Kant claims that "their representation is an act of the subject itself and a product of the imagination for the sense of the subject."

86 KrV §24 B152 had "*Wirkung* des Verstandes auf die Sinnlichkeit"; and B154 had "*Einfluß* des Verstandes auf den inneren Sinn."

87 In this I disagree with Kitcher (1986, 137) who reads the argument of the Fn. as tying the unity of spatial and temporal representations to concepts (and manifests surprise at the Fn.'s ending, which assigns the unity to space and time as intuitions). In this text, Kant consistently differentiates intuitions—both the forms of intuition and the formal intuitions—from concepts. Kitcher's confusion, however, is warranted if the imagination is counted as a function of the understanding.

ity" before actually forming any judgment (for no concept is available yet), it is the productive imagination that properly "determines sensibility" for the understanding's judgment (or determines itself as receptivity, hence *gives* the intuition). The spontaneity of the imagination makes the receptivity of sensibility possible, which is what the understanding in fact determines. Thus, space and time as intuitions "are given" only if the imagination provides the *synthesis speciosa* through which first the understanding is able to form judgments, i.e., to determine sensibility. And since the *synthesis speciosa* is the condition through which "space and time are first *given* as intuition, the unity of this a priori intuition belongs to space and time, and not to the concept of the understanding."[88] Thereby Kant makes two points: first, it now becomes retrospectively clear that the space and time presented in the Transcendental Aesthetic are truly the product of the synthesis of imagination—such synthesis is required for *both* formal intuitions (which present "unity of representation") *and* forms of intuition (receptivity itself requires the activity of imagination's spontaneity);[89] and second, the unity produced by the imagination belongs a priori to sensibility (determining its very receptivity), and not to the understanding.

Thus, with regard to the claim presented at the beginning, we can now conclude that the novelty of Kant's theory of sensibility is not only the discovery of an a priori formal component essential to the cognitive process, which lies next to the passive and material aspect generally recognized by pre-Kantian epistemological and psychological theories. Once the imagination is introduced as the crucial active factor of sensibility productive of the *synthesis speciosa*, the relation between the two aspects of sensibility comes to the fore. The spontaneity of imagination makes sensibility as receptivity possible—even the givenness of the manifold of the forms of intuition rests on the more original unity brought about by the imagination's synthesis.

4 Kant's Imagination. A Mediating Role?

In *Glauben und Wissen* Hegel gives his appraisal of the relation between the principle of the synthetic unity of apperception, which emerges in the deduction, and the 'isolated' presentation of space and time in the Aesthetic, recognizing the more advanced view of sensibility offered by the notion of a synthetic and productive activity of the imagination. Hegel sees in the "productive imagination"

---

88 KrV B160f. Fn.
89 See Longuenesse (2000), 217 ff. for a defense of this interpretation.

the highest "principle of sensibility," clearly set against Kant's own description of sensibility as mere receptivity and itself a moment of what, for Hegel, the speculative concept of reason is. While the "true idea of reason" (namely, the idea of the "absolute identity" of the "*Ungleichartiges*") is contained, in Hegel's view, in the chief question of the first *Critique* concerning the possibility of synthetic a priori judgments,[90] he maintains that in Kant's presentation "one catches glimpses of this idea in the shallowness of the deduction and, in relation to space and time, not there, where it ought to be, namely, in the transcendental exposition of these forms [i.e., in the Transcendental Aesthetic], but only in the sequel, where the original synthetic unity of apperception comes to the fore in the deduction of the categories and is also recognized as principle of the figurative synthesis, i.e., of the forms of intuition. Here space and time are themselves conceived as synthetic unities, and the productive imagination, spontaneity, and absolute synthetic activity are conceived as the principle of sensibility, which heretofore has been characterized only as receptivity."[91] On Hegel's account, the idea of reason as the synthesis and unity of the heterogeneous elements of sensibility and understanding is rooted in sensibility itself and finds in the synthetic and spontaneous (yet "blind") activity of the imagination its highest representative: space and time are themselves synthetic unities which reproduce the idea of reason. Ultimately, the productive imagination is intuition synthesized; it is the power that through its peculiar synthesis (the *synthesis speciosa*) yields the space and time that were the topic of the Transcendental Aesthetic. Coming somehow close to Heidegger's position, Hegel sees in the original synthetic unity of the apperception-productive imagination a "unity that should not be conceived as the product of opposites but rather as the truly necessary, absolute, original identity of opposites."[92] On this basis, however, Hegel easily moves on to erase Kant's radical separation between sensibility and understanding placing the imagination above both (and within both) as their original unity. He then concludes that "one and the same synthetic unity [...] is the principle of intuiting and of the understanding."[93] Moreover, the mediation achieved by the dialectically reinterpreted unity of the imagination does not imply the emergence of a third term. The Kantian imagination, Hegel contends, should not be taken "as the middle term (*Mittelglied*) that gets inserted between an existing absolute subject and an absolute existing world; it must rather be recognized as what is primary and original and as that out of which the subjective I and the objective

---

90 *Glauben und Wissen*, GW 4, 326 f.
91 *Glauben und Wissen*, GW 4, 327 (the passage is cited also by Waxman, 1993, 74).
92 *Glauben und Wissen*, GW 4, 327.
93 *Glauben und Wissen*, GW 4, 327.

world first sunder themselves."[94] In this case, mediation, within the unity of reason's idea, is indeed seen as overcoming the dualism of sensibility and understanding. It does so, however, by mobilizing a (dialectic) concept of identity (of opposites) and a (speculative) notion of reason that are entirely Hegelian but hardly compatible with Kant's position in which the problem of the imagination is rather the problem of *synthesis*.

As shown above, Hegel's insight that the concept of the productive imagination and its synthesis implies a re-reading of the Transcendental Aesthetic and leads Kant to a more complex theory of sensibility is indeed correct. Now I want to argue, albeit only briefly and by way of conclusion, against Hegel's view that the synthesis of the imagination is the unitary principle of intuiting and the understanding, and claim that the function that the imagination plays in the B deduction does not undermine the separation of *Sinnlichkeit* and *Verstand*, but actually confirms it. Thereby, I shall bring my argument in favor of the imagination's belonging to the sensible faculty to bear on the question of its alleged 'mediating' function.[95] Whereas is the passage discussed above, Kant does not use the language of mediation to describe the imagination's activity, there is certainly a sense in which what is at stake in the deduction is the problem of bridging the gap between the sensible and the intellectual—or mediating, as it were, the blindness of both intuition and the imagination on the one hand, and the emptiness of concepts on the other.[96] However, while (Hegelian) mediation implies the concept of a unity that reconfigures the relation of the two heterogeneous (and opposed) extremes, and shows, at the same time, that their unity precedes and is more original than their distinction, for Kant sensibility, sensibility and understanding remain (and constitutively are) two distinct branches of human knowledge throughout, even—and, I suggest, even more strongly—when the productive imagination and its *synthesis speciosa* is introduced as that which gives the concept a determinate intuition. In this case, Kant insists on the fact that our human understanding is not itself intuitive, i.e., that it needs the imagination to "represent in intuition" the sensible given. This is the burden that the understanding puts on sensibility—the *Wirkung* that it exercises on it.

---

94 *Glauben und Wissen*, GW 4, 329.
95 The discussion of this last question should include an account of the schematism, which I cannot offer in this essay but which I plan to pursue in a second essay.
96 See, in very different perspectives, Sallis (2000, 66); Waxman (1993, 75).

Thus, if we do want to claim that the imagination plays a mediating function or the function of a "bridge" connecting two heterogeneous terms,[97] mediation should be understood neither in a Hegelian way, i.e., as implying the idea of an original unity from which the separation of sensibility and understanding issues, nor as requiring the intervention of a third term capable of connecting the two extremes. The bridging function of the imagination is nothing more than the very dynamic movement or activity whereby the transition between the two takes place. The imagination does not 'mediate' between sensibility and understanding as if it were either a third term placed between the two or their original and unitary source—the imagination is the very movement of the embodied mind as it dynamically utilizes the two heterogeneous branches of its cognitive activity for actual cognition. If any mediation is involved in the imagination's activity, it is rather that which takes place between two aspects of sensibility itself, namely, between receptivity and spontaneity or productivity. The imagination shows that, transcendentally, sensibility's spontaneity comes before its receptivity (the unity of intuition comes before the given manifold in it, hence before the sensible given) and makes it possible.[98] This, however, is hardly mediation. Properly, what the imagination as an utterly sensible faculty does is reconfigure, in the advanced transcendental perspective of the Analytic, the theory of sensibility of the Aesthetic. At this level it shows that within the process of cognition guided 'from above', so to speak, by the understanding as the faculty to form judgments, sensibility's spontaneity comes before its receptivity, intuition before the object—and before the concept.

---

97 A "bridge" quite in the sense advanced by Kant in §IX of the Introduction to the KU (AA V, 36): the two domains of nature and freedom remain, in this case, separate; they are connected, however, through a "bridge." Unlike the faculty of judgment, however, the imagination is not a third, separate faculty. And yet there is a sense in which *Urteilskraft* is both a third faculty (because it possesses an a priori principle of its own) and a pervasive function at work in relation with the understanding (and belonging to it). The imagination is not a third faculty as it is not endowed with an a priori principle of its own. It belongs to sensibility and displays its bridging function coming from this side of the divide.

98 This offers, in my view, Kant's solution for a connected problem raised by the literature, namely, the difficulty of talking of sensibility as mere receptivity and, at the same time, as endowed with an activity of its own while not reducing it to the "noetic intuition" of Plato and Aristotle, i.e., while not claiming that the matter which affects our senses is something that our spontaneity has always already formed (see Ferrarin, 1995, 159). The thesis of transcendental idealism, hence the preserved radical separation of sensibility and understanding, further reinforced by the fact that the imagination is reclaimed as a sensible faculty are Kant's answer to that problem.

References

Descartes, R. (1897–1910, 1996): *Oeuvres complètes*, ed. Charles Adam, Paul Tannery, Paris: Vrin. (=AT).

Hobbes, Th. (1968): *Leviathan*, ed. C.B. Macpherson, Harmondworth/NY: Penguin.

Hegel, G.W.F. (1968 ff.): *Gesammelte Werke*. In Verbindung mit der Deutschen Forschungsgemeinschaft hrsg, v. der Reinisch-Westfälischen Akademie der Wissenschaften. Hamburg: Meiner (=GW).

Kant, I. (1910 ff.): *Kants gesammelte Schriften*, hrsg. v. Der Preußischen Akademie der Wissenschaften, Berlin: De Gruyter. (=AA).

Allison, H. (1983): *Kant's Transcendental Idealism: An Interpretation and Defense*. New Haven: Yale University Press.

Baum, M. (1986): *Deduktion und Beweis in Kants transzendentalen Denduktion*. Königstein: Hain Verlag.

Ferrarin, A. (1995a): "Kant on the Exhibition of a Concept in Intuition." *Kant Studien*, 86, 131–174.

Ferrarin, A. (1995b): "Kant's Productive Imagination and its Alleged Antecedents." *The Graduate Faculty Philosophy Journal*, 18, 1, 1–27.

Ferraris, M. (1996). *L'immaginazione*. Bologna: Il Mulino.

Frede, D. (1992): "The Cognitive Role of *Phantasia* in Aristotle." Nussbaum, M./Oksenberg Rorty A. (eds.), *Essays in Aristotle's De Anima*. Oxford: Clarendon Press, 279–295.

Heidegger, M. (1973): *Kant und das Problem der Metaphysik*. Frankfurt a.M.: Klostermann (tr. by R. Taft, 1990. *Kant and the Problem of Metaphysics*. Bloomington: Indiana University Press).

Henrich, D. (1955): "Über die Einheit der Subjektivität." *Philosophische Rundschau*, 47, 3, 62–69.

Kitcher, P. (1986): "Connecting Intuitions and Concepts at B160n." *The Southern Journal of Philosophy*, 25, Supplement, 137–149.

Kneller, J. (2007): *Kant and the Power of Imagination*. Cambridge: Cambridge University Press.

Long, Christopher P. (1998): "Two Powers, One Ability: The Understanding and Imagination in Kant's Critical Philosophy." *The Southern Journal of Philosophy*, 36, 233–253.

Lyons, John D. (1999): "Descartes and Modern Imagination." *Philosophy and Literature*, 23, 2, 302–312.

Longuenesse, B. (2000): *Kant and the Capacity to Judge*. Princeton: Princeton University Press.

Makkreel, R. (1990): *Imagination and Interpretation in Kant*. Chicago: University of Chicago Press.

Mörchen, H. (1930): "Die Einbildungskraft bei Kant." *Jahrbuch für Philosophie u. phänomenologische Forschung*, 11, 311–495.

Nuzzo, A. (2008): *Ideal Embodiment. Kant's Theory of Sensibility*. Bloomington: Indiana University Press.

Nuzzo, A. (2005): *Kant and the Unity of Reason*. West Lafayette: Purdue University Press.

Sallis, J. (2000): *Force of Imagination. The Sense of the Elemental*. Bloomington: Indiana University Press.

Sepper, D. (1989): "Descartes and the Eclipse of Imagination, 1618–1630." *Journal of the History of Philosophy*, 27, 379–403.

Sepper, D. (1988): "Imagination, Phantasms, and the Making of Hobbesian and Cartesian Science." *The Monist*, 71, 526–542.

Waxman, W. (1993): "What Are Kant's Analogies About?" *The Review of Metaphysics*, 47, 1, 63–113.

Waxman, W. (1991): *Kant's Model of the Mind. A New Interpretation of Transcendental Idealism*. NY/Oxford: Oxford University Press.

Christian Helmut Wenzel
Art and Imagination in Mathematics

Discovery or Invention?

In the chapter "Deduction of Pure Aesthetic Judgments" from his *Third Critique*, Kant develops a theory of genius (sections 45–50) which leads him to the claim that there can be genius in the arts but not in the sciences. For instance in section 47 he argues that someone like Newton can be called "a great mind" (*ein großer Kopf*) but should not be called a "genius" (*Genie*). The reason he gives for this is that everything Newton has discovered can be learned, whereas one cannot learn how to write poetry. Newton can show us step by step what he has done, but Homer and Wieland cannot. Kant writes:

> Thus everything that Newton expounded in his immortal work on the principles of natural philosophy, no matter how great a mind it took to discover it, can still be learned; but one cannot learn to write inspired poetry, however exhaustive all the rules for the art of poetry and however excellent the models for it may be. The reason is that Newton could make all the steps that he had to take, from the first elements of geometry to his great and profound discoveries, entirely intuitive not only to himself but also to everyone else, and thus set them out for posterity quite determinately; but no Homer or Wieland can indicate how his ideas, which are fantastic and yet at the same time rich in thought, arise and come together in his head, because he himself does not know it and thus cannot teach it to anyone else either. In the scientific sphere, therefore, the greatest discoverer differs only in degree from the most hard working imitator and apprentice, whereas he differs in kind from someone who is gifted by nature for beautiful art. (AA 05: 308–9, section 47)[1]

In this picture we have discovery and learning in the sciences and the arising of ideas in the arts. There is a strict separation. But I wonder whether Kant went too far in drawing this distinction and by excluding Newton from the sphere of genius. On the one hand, Kant is right in pointing out that there are definitions, theories, and proofs in the sciences which allow us to learn those sciences step by step, while this cannot be said of the arts. There are no proofs in matters of art. There is no deductive reasoning. But on the other hand, not just anyone can discover definitions, theories, and proofs, and we often say, contrary to what Kant claims, that mathematicians and scientists are geniuses. We say Einstein was a genius. Understanding Einstein's theory and following a proof step by step is one thing (and difficult enough), but discovering the theory and finding the

---

1 References are to volume and page numbers of the *Academie Ausgabe* (AA).

proof oneself is an altogether different matter. Not everyone can be said to be capable of having done what Einstein and Newton did. Why should discovery not be a sign of genius, even if later on others can learn the theory and follow a proof step by step?

Kant writes that poets cannot say how their ideas "arise and come together in their heads". But is this so different in the case of scientists? They have ideas, too, and they usually do not arrive at them in purely deductive ways.[2]

But according to Kant, "even if one thinks and writes for himself, and does not merely take up what others have thought, indeed even if he invents a great deal for art and science, this is still not a proper reason for calling such a great mind [...] a genius" (AA 05: 308, section 47). For Kant, the reason for this is that in the sciences there is some kind of "natural path" of discovery, whereas in the arts there is no such path. Thus he goes on to say that "just this sort of thing could also have been learned and thus still lies on *the natural path of inquiry and reflection in accordance with rules*" (AA 05: 308, section 47, italics mine). I have doubts about this, and in the following I will repeatedly come back to this point.

It is true that there is objectivity and progress in the sciences in ways that cannot be found in the arts and that one is therefore led to speak of a "natural path of inquiry and reflection in accordance with rules" (*auf dem natürlichen Wege des Forschens und Nachdenkens nach Regeln*). Kant even thinks that this path is unique, as if there were no other one possible. He writes "*auf dem natürlichen Weg*" and not "*auf einem natürlichen Weg*". For him there is only one. But I wonder whether this path is indeed that "natural" and unique as Kant thinks it is. Contrary to what Kant claims, it seems to me that when we move into more abstract mathematics, there is much freedom in how we can set up definitions and even whole theories. Alternative theories and alternative proofs are possible. Starting from the natural numbers, we can move to the rational, the real, and the complex numbers, each extending the previous one. But we can also, alternatively, move into very different kinds of number systems. Starting from the nat-

---

2 Of course I am not the first one to find Kant's claims and arguments not convincing here. Thus Donald W. Crawford already writes: "I have no confidence that mathematicians can tell us how they know how to begin an unobvious proof or construction any better than Homer and Wieland could have told us how their ideas came together in their heads" (Crawford 1982, p. 166). He criticizes Kant's arguments by observing: "And this surely is a confusion between the order of discovery (*ordo inveniendi* or *ordo cognoscendi*) and the order of teaching or systematic exposition of truth already discovered (*ordo docendi*)." (p. 165) Already Aristotle was aware of the fact that proofs are often written down (deductively) in the opposite way compared to how they were discovered (inductively and intuitively).

ural numbers, instead of adding new numbers, we can systematically identify certain numbers with each other. We can choose a prime number, say 7, and identify each number n with the numbers $n+7$, $n+14$, $n+21$, $n+28$, etc. This is an alternative. From here you cannot extend to the complex numbers any more. You either go this way, or you go the other way.

We can also adopt abstract axiomatic theories of so-called number "fields", which includes the complex numbers as well as the strange case involving the number 7 indicated above. Here we can have "algebraic" numbers of "positive characteristics". In such fields it might happen that adding the unit 1 several times to itself will give zero (as in the case with the number 7): $1+1+...+1=0$, or (if we loosen the requirements for a field a little) that usual and commonsensical rules such as $ab=ba$ or $a(bc)=(ab)c$ no longer hold. These are strange "numbers". It took mathematicians some time to accept what we now call the "complex numbers" as the "natural" extension of the real numbers. I think there are good reasons for saying that there is not only discovery but also invention involved in mathematics. Even if we say that mathematics is already out there in a Platonic realm of ideas, or in a Fregean third realm of thoughts, waiting to be discovered, I think it matters what we see and cut out from this realm. This is an idea I want to explore in this essay. It simply matters to us (what theory we have), and it seems to me that Kant is downplaying, or not sufficiently aware of this part: the seeing and cutting. This is more than discovery. It involves intuition and making choices.

Differential and algebraic geometry are very complicated and rich theories that lead to many questions which we have not answered yet, and it does not seem obvious to me that there is only one way in which these theories will be developed in the future. Regarding physics, I wonder whether relativity and quantum theory lie on "the natural path". Nor is it clear to me that they do so "in accordance with rules". What exactly should those rules be? Would they be methodological meta-rules of enquiry? Would they be rules of mathematics proper? Are there rules that tell us which rules we should adopt? What are the rules that make us think of a "natural path", or even "the" natural path?

Usually it is the case that rules surface only *post factum*, such as the axioms of what we now call "fields" and of which the real and the complex numbers are special cases. Who knows what else is swimming under that surface? Is there only one, single path of investigation and development? How about statistics, probability, methods of approximation, and the use of computers? How much of which field of mathematics will be developed and practiced? What will surface? Everything? Social, economic, political, and many other factors that are contingent from the point of view of mathematics proper come into play here. There are different styles of practicing mathematics, and different styles make for different paths. Different paths lead to different theories, and different theories are the objects for the next generation of researchers to work on. It seems to me there is no single path, neither regarding method nor regarding result. We are not aware of what has not surfaced and what we left at the side. There are choices involved in the production and construction of theories, and these choices will be imbedded in those theories even if they will be forgotten as such. We often get used to what we have and then think of it as the only possibility. But sets of accepted ways of reasoning keep changing. So do sets of accepted proofs, practices, styles, and the questions mathematical communities consider worth asking (see Kitcher, pp. 149–271).

One basic difference between the arts and the sciences is that in the arts we are dealing with individual works, individual poems, paintings, and musical compositions, whereas in the sciences we are dealing with general and abstract rules and theories. A poem can be repeatedly read and memorized, a painting can be looked at, and a musical composition can be performed and listened to. Works of art are objects of the senses, whereas a scientific theory is something abstract that is applied and not contemplated, so one could argue. It is this aspect of applicability that makes the scientific theory so useful and powerful, one might say. But for poems, paintings, and musical compositions there are no such applications. They are individual works of art to be cherished for what they are in themselves. They are objects of the senses and are supposed to speak for themselves. Thus individuality, sensibility, empirical factors, and contingency play a role in the arts and not in mathematics, so it seems. Mathematics is universal and a priori. I do not want to challenge this universality and a priority. But I wish to challenge the idea that it is only those very objects we have in our mathematical theories now that are possible. I want to show that there are other mathematical objects and theories that are possible as well, but have not been discovered and maybe never will be talked about. There are other "natural" paths.

If we do not think about aspects of application and focus instead on the scientific theories themselves, as ends sought for their own sake (as scientists often do), the distinction between the arts and the sciences becomes blurred. Pure mathematics is usually done for its own sake, comparable to how art is done for art's sake, *l'art pour l'art*. A mathematician can work and live in his or her world of abstract ideas and problems, comparable to how Vincent van Gogh lived in his world of colors, shades, and shapes and light in nature and in his paintings. Both see the world in unique and individual ways, and it will take time for others to see what they saw. Mathematicians and painters can set new standards. Both can be diligent and work to exhaustion and even insanity. The great discoveries in the sciences were *not* completely guided by rules. Contrary to what Kant claims, they were *not* simply lying on "the natural path of inquiry and reflection in accordance with rules". I think there is no such path.

There are many different possible criteria of what we want, or might want, criteria of usefulness that depend – in addition to criteria of truth – on aesthetic, economic, political, and other factors. These factors do not determine what is true, but they affect our interests. They affect where we look and what we keep and value as part of our new theories. What we discover depends on our choices regarding which axioms to adopt and which theories to develop. If we go this way, we will discover X. If we go that way we will discover Y. Once you embark on X, you might never see Y. One might say that this does not matter, because mathematical theories are already out there, in a Platonic realm, waiting to be discovered. But then I could say the same about all poems, written and unwritten. I could say that they too existed already in some Platonic realm. Hence discoveries in mathematics are not completely different from creations of works of art. They are creations and inventions, too, as the latter can also be seen as discoveries.

It thus seems to me that the difference between mathematics and the arts is not as radical as Kant depicts it to be. Let us look at the idea of extension and development again. Einstein broke with Newtonian physics. He was guided by some classical ideas (indeed "in accordance with rules"), but he also did not follow all of them. He broke some of those rules (*not* "in accordance with rules", contrary to what Kant claimed). Here mathematics can be compared with the arts. Artists work within traditions, but they also break with these traditions. They create and initiate new movements. Kant says that "no Homer or Wieland can indicate how his ideas ... arise and come together in his head, because he himself does not know it and thus cannot teach it to anyone else either". But could Newton or Einstein do this? Could they "indicate" much better than Homer and Wieland how they arrived at their ideas in the natural sciences? Better yes, but completely? Does discovery not take inspiration, too? Does it not take *"Einfälle"*, as one says in German, that is, sudden "intrusions" and "impacts" from the outside, from outside the field of study itself? This aspect of an "outside" is, I think, what contradicts Kant's idea of a "natural path". The outside can later become part of the inside, part of the mathematics we now have. What was accidental has led to something that now seems natural and the only possible way.

It is true that once we have familiarized ourselves with the definitions and basic theorems of a theory or practice, we will be able to follow a proof step by step. But even then we cannot immediately see, at each step of the proof, why we should do this and not that. Following is not the same as understanding and doing it oneself. Let us look more closely at the differences. Why should we at this point in the proof apply this theorem and not another? Why should we, at this point, set $x=3$ and not $x=5$ or 119 or any other number? In the course of a single proof many choices are made, choices among already established theorems and choices among infinitely many possible applications (what to apply a theorem to). These choices are not dictated by the definitions and theorems themselves. That is why proofs are so hard to find. I think it is for this reason that also Newton and Einstein cannot teach us what they did and how their ideas "arise and come together" in their heads. They themselves do not know it.[3]

Only when we see the result of the proof as a whole, at the end and in retrospect, after having gone through the proof many times, will it appear to us as if it were all natural and lying along a "natural path". Only then will it seem to us as if there were only one possible path. But this is wrong. Later on, others will find shorter proofs. They will find *other* "natural paths" that lead to the same result or a more general one. If the latter happens, the whole picture will change. This might even lead us to change some of our assumptions. A change of perspective can go all the way down to the basics. To put it into more general terms: The mind affects what we see – not the things themselves, but their appearances (and this is all we have). The fact that certain things appear and others don't has to do with us.

I think Kant could have stressed and further developed the aspect of originality and creativity in mathematics by referring to his analytic-synthetic distinction and his view that mathematical propositions are synthetic and not analytic. Synthesis allows for imagination and intuition to play more substantial roles. But Kant did not do this. He did not go this way. For him, when it comes to math-

---

3 Henri Poincaré (1854 – 1912) defended the relevance of invention and intuition in mathematics. As there are logical-analytical minds (*analyste*), so there are geometric-synthetic-intuitive minds (*géomètre*), he says. Knowing the rules is not enough, and this is similar to how we play chess: "*de tous ces chemins, quel est celui qui nous mènera le plus promptement au but? Qui nous dira lequel il faut choisir? Il nous faut une faculté qui nous fasse voir le but de loin, et, cette faculté, c'est l'intuition. Elle est nécessaire à l'explorateur pour choisir sa route, elle ne l'est pas moins à celui qui marche sur ses traces et qui veut savoir pourquoi il l'a choisie*" (p. 36); "*l'intuition est l'instrument de l'invention*" (p. 37). It is the end that justifies the "why" (*pourquoi*). Poincaré is sympathetic to Kant's notion of intuition (*Anschauung*) as well as his idea of the synthetic *a priori*. Also Charles Parsons has argued for the necessity of Kantian intuition in mathematics. But Parsons focuses more on the role that intuition (*Anschauung*) plays in our *constructing* mathematical objects (following the idea that concepts without intuition are blind) and he focuses less on the aspect of intuition in the choices and value judgments we make (intuitive discovery, German *Intuition*, not German *Anschauung*). Charles Parsons focuses more on the object, whereas I here focus more on the method and the fact that different methods lead to different objects.

ematics, intuition and imagination were merely subservient to rules of the understanding. Although we need intuition and imagination to carry out proofs, they do not contribute something on their own. It is this last point that I wish to question.

Let us return to the distinction between singularity and generality. The arts give us single, individual works of art, and the sciences offer general theories. Art is appreciated and contemplated, whereas the sciences and their theories are applied. For Kant, mathematical and scientific theories consist of rules. At one point he even says that "mathematics is nothing but rules" (*Reflection* 922, AA 15: 401). Once we understand the rules, we can apply them and they seem to determine everything. In Kant's picture, they put us on firm and fixed rails, whereas works of art do not function in this way. Indeed, instead of speaking of applications, we say that works of art "speak for themselves" and unfold worlds of their own. They are full of surprises. Kant says artists produce "models" (*Muster*) which other artists imitate (*Nachahmung*, AA 05: 309) in order to become original artists themselves. They do not just copy (*Nachmachung*, AA 05: 309). Works of art are exemplars and not rules, even if they seem to be "an example of a universal rule that we are unable to state". But I think this distinction can become blurred when we look more closely. On the one hand, students of mathematics and physics can try to understand scientific theories by trying to find proofs themselves. They can think ahead and then come up with new proofs. They do not blindly and mechanically apply old rules. In doing this they are usually guided by their favorite examples and their individual interests and backgrounds. Thus the individual perspectives of mathematicians do matter to what is and what is not discovered. On the other hand, art sometimes is mechanically reproduced and applied, for instance in advertisement and products of mass-consumption, for instance CDs, perfumes, posters, clothes, cars, motorcycles, etc. The line between mechanical *Nachmachung* and creative *Nachahmung* in the arts is not clear-cut. And there is a similarly foggy line within mathematics, namely in our practicing, learning, and doing research in mathematics. How much do we understand ourselves and to what degree are we just blindly following others? If we say we do it "ourselves", who do we think we are? Do we ever really do anything all by ourselves? Wittgenstein has shown us some of the intricacies of such questions in his reflections about drill in mathematics (although I think he went too far).[4]

---

4 I think he went too far in reducing rule-following to drill and habits, and he went too far in reducing normativity to social and evolutionary factors (see Wenzel 2011, "On Wittgenstein on Certainty"). But how to read Wittgenstein is disputed. Some scholars do not read him as making or suggesting such reductions.

It should by now be clear that there is something problematic about saying that mathematics is "nothing but rules". Let me focus on the aspects of particularity, individuality, and singularity again. A mathematician who does research often does not work so much with abstract rules but instead has a typical example in mind. Such an example serves as a model that allows him to see certain essential features concretely, so that he can manipulate and modify these features to better suit the problem he is trying to solve. If one understands one or two cases very well, one has a good handle on the whole theory. Mathematicians who do research know this well. One often goes by examples when doing research, because there one can "see" what is going on when one changes some parameters or some basic assumptions while keeping others fixed. One can see what will happen if one twists those exemplary models a little here or there to suit or challenge the problem at hand. In this way such models can serve as "models of imitation" and models of modification, similar to exemplars in art. A crucial point in this is their concreteness and particularity, because *it is this concreteness that allows one to see more than what is implied by the general and abstract rules alone.*

I do exactly that when I work for instance on linear algebraic groups in general by having the particular group SL_3 in mind. The group SL_3 has properties a group in general does not have. It is by focusing on the particular example and model SL_3 in its concreteness (which is more specific than the theory in general and thereby goes beyond that theory) that new ideas and insights often arise. Aspects from "outside" (not within the general theory) thus can come in. This can even lead one to change some of the rules one started out with. One can change axioms or assumptions of the framework, or even change the framework itself. In that sense researchers often work like artists. They sometimes follow concrete models and not abstract rules. Thus Kant's own theory of genius and the arts could have led him to see such aspects of creativity in mathematics, had he only paid more attention to what is actually going on in learning and doing research in mathematics. I think it would have been more obvious to him, had he been more interested in the higher mathematics of his day, such as infinitesimal calculus, as Hegel was, or in learning mathematics as Wittgenstein was interested in. It will be more difficult (but not impossible) to see the aspects I want to bring out, if one restricts one's attention to triangles and *5+7=12* as Kant more or less did.

Kant's exclusion of scientists from the realm of geniuses (AA: 05:308 – 9, section 47) occurred in the development of his general theory of genius. He famously says, "Genius is the talent (natural gift) that gives the rule to art. [...] Genius is the inborn predisposition of the mind (*ingenium*) through which nature gives the rule to art" (AA: 05:307, section 46). He thinks that it is "through" Homer and his genius that nature gives the rules of epic poetry to art, whereas Newton did it himself, through his own diligence, and we all could do it if we were only diligent enough. Homer therefore cannot explain what he did, but Newton can. Kant of course has a point. There is more freedom in the composition of a poem than in the development of a proof. One can change words here and there and still have an epic poem, whereas if one changes formulas in a proof one might end up not having any proof of anything anymore. The proof will suddenly not work and be invalid, having zero value. Mathematical proofs are strict in that sense.

But in spite of this strictness, proofs can also change. If a proof is as long as the *Odyssey*, one will easily get lost and reading and following will become an odyssey itself. Later a shorter proof will usually be found, and then one sees that there was more than one way "the" proof could go. Newton will therefore be hard pressed to explain every step. He actually cannot. The reason is that there is actually *no necessity* in those steps. There is no necessity in why *this* theorem is applied here and not another, why it is applied in *this* way and not another, why I refer back to *that* previous result and not another. It is I – the reader, or whoever carries out the proof – who does the choosing and referring. *The necessity is only in the "if-then", not in the "if" part. Nor is there any necessity in where in the proof the "if" part occurs.* If you apply this theorem here in this way, such and such will necessarily follow. But if you do not apply it, nothing will happen; and if you apply another theorem, something else will happen. Other proofs are always possible. Even the whole Newtonian theory, the framework within which a proof by Newton is carried out, is not the only possible one. Leibniz had another. There is analysis and there is non-standard analysis. There is Euclidean geometry, and there are non-Euclidean geometries. There are fields of characteristic zero, and there are fields of positive characteristic p. There is plenty to choose from. There are alternatives. It is not a one-way road. Sometimes things are going the other way, and there are junctions. Even within a fixed framework there is much to explore and much to choose.

Thus, contrary to Kant, there is no unique and "natural path of inquiry" (309). There is lack of necessity at least at two levels, regarding theories chosen and regarding steps taken within those theories. Theories keep developing and changing. Leibniz offered an alternative way, and Weierstraß, Riemann, and Einstein later on had altogether new ideas. Kant talks of reflection and inquiry "in accordance with rules", but it seems to me these rules are not prescribed. They did not fall from the heavens. Peano said that the natural numbers are given by God, but the rest is done by us (as far as mathematics is concerned). Higher mathematics is more than just the set of natural numbers with addition. The rules of any advanced mathematical theory are not as predetermined as Kant takes them to be. We can change them and pick others. Even if mathematics appears to be "nothing but rules" in the "if-then" sense, we can still ask: Which rules? And when and where are they used for what? Hence we can ask: Which mathematics? It seems to me that Kant's arguments for the exclusion of mathematics and the sciences from the domain of genius do not sufficiently take into consideration our actually doing mathematics.

This is not only a psychological factor. Our doing research has lead to the mathematics we now have, and we could have developed another. I don't want to say that *2+3* could be *0* in our ordinary sense of natural numbers. But if you work over fields in characteristic *5*, this is what will happen. There you write and think *2+3=0*. If you work with the real numbers, you take it for granted that *ab=ba*; but if you work with matrices, you will not. The rules depend on objects and theories, and it is our choice what objects and theories to work with. There are undiscovered theories and roads not taken, just as there are poems not written. Although mathematics is not the same as poetry or painting, and although there are differences, it seems to me these differences are more a matter of degree and not as absolute as Kant thought.

Imagination

Kant claimed that mathematics is "nothing but rules" and (for this reason I think) he excluded it from the realm of genius and beauty. Nevertheless, I think Kant had all the tools at his disposal to do otherwise. He had a rich theory of schematism and imagination, which would have allowed him to explain aspects not only of construction, but also of invention, genius, and beauty in mathematics. In addition to this, his regarding mathematics as essentially synthetic would have invited such explanations. This applies especially to our ways of actually doing mathematics, learning, and doing research in higher mathematics. In the following I will try to use some of Kant's insights into the nature of imagination and some of his ideas about schematism to reveal such aspects.

In the first *Critique* sensibility and understanding are established as the two pillars of human cognition. They give us intuitions and concepts. How exactly they depend on each other and "cooperate" has been a question of discussion and controversy ever since Kant. According to him, intuitions without concepts are "blind". But if we want to say that infants and animals are not blind, should we then have to say that they already have concepts, or at least the categories? Infants develop conceptual capacities, but most animals never do. Does Kant's conception of 'transcendental synthetic unity of apperception' apply to them? And, to approach the question from the other end (on the line from animals to infants and fully grown up humans), do we adult human beings always think when we perceive? Do we need concepts and language in our every-day perceptions? Or is there some kind of non-conceptual, pre-conceptual, or pre-predicative perception, some kind of *vorprädikative Wahrnehmung* (Husserl)? Is there some kind of "simple seeing" (Fred Dretske – to name just one analytic philosopher who takes this view)? Is there some pre-conceptual but nevertheless meaningful and not blind experience when we don't pay attention, day dream, doze off, or gradually wake up in the morning not knowing whether we are awake or dreaming (Hermann Schmitz)? These are questions that have been much discussed in attempts to clarify the interplay between perception and conception in the light of theories of vision, evolution, and cognitive science. They are discussed in our trying to specify what is distinctive about us human beings in comparison with other, non-human animals.

Are we just another kind of animal? Or are we special creatures that should not be called "animals" at all? In the Greek tradition we are animals that have "logos", we are ζωα λογον εχοντα, *animalia rationalia*, rational animals. But what exactly "rationality" is supposed to be is disputed. Biologically there seems to be continuity through and throughout evolution. But biology offers only one perspective. Linguistics, anthropology, and philosophy offer other views.

In Kant's philosophy, imagination serves as a link between sensibility and understanding. The categories of the understanding are schematized in modes of time and space in their applications to sensible intuitions. Kant is interested in justification, not in genesis and development. But a separating line is still difficult to draw. He saw the relevance of imagination in his theory of threefold synthesis and in his deduction of the categories in connection with consciousness and the unity of apperception. But the line between empirical and transcendental aspects is not clear. Nor is the role of imagination. How "spontaneous" and how independent imagination is from the power of the understanding is a delicate question. How free is imagination from rules? What exactly are rules? These questions are not idiosyncratic to the Kantian system. When suitably translated, they arise in current philosophical discussions of cognitive science and theories of vision. They appear also in current discussions of internalism versus externalism in regard to meaning and perceptual content. John McDowell claims that concepts go "all the way out", contrary to Gareth Evan's views (see Wenzel 2005). Hilary Putnam's doubts that meanings are "in the head", and his arguments have been extended by Tyler Burge to issues about perception and perceptual content. Meaning involves not only concepts but also the environment, physical and social. Perception thus involves capacities comparable to Kantian intuition and imagination. The terminology is new, but many of the problems are old.

The general question of how independent imagination is from the understanding can be asked particularly with respect to mathematical objects. We need intuition and imagination to draw lines in space. We need them to carry out proofs in geometry, and we also need them to conceive of numbers in successive synthesis in arithmetic (see the work of Charles Parsons). For Kant the two sides of the equation *5+7=12* are different due to their intensions, not their extensions (Iseli p. 90). They are different in content, because they were arrived at in different constructive ways, and these ways require intuition and imagination (Wenzel 2011, "Urteil" p. 2288). The equation is true for synthetic and not analytic reasons. Kant's views are non-standard, even for his own time, and they depend on his understanding of time and space as being "subjective" in the framework of his transcendental philosophy. Frege distinguished between "sense" (*Sinn*) and "reference" (*Bedeutung*) to explain the truth and nontriviality of an equation such as *5+7=12*. But for him it was more reference and extension that matter, and less sense or intension (not to speak of intention). Even though he speaks of "sense", his notion is similar to the notion of reference. Husserl held on to another aspect of "sense". For Frege, sense is purely objective, to be found in the third realm, timeless and to be grasped. Sense is not made. He does not pursue a theory similar to Husserl's intentionality and *noema*, and he does not accept the Kantian notion of intuition or the framework of transcendental philosophy either.

In the third *Critique*, imagination takes a much more central role than in the first, which is no wonder because judgments of taste allow imagination to be free from rules of the understanding and to make substantial contributions (in the free play of imagination and understanding and the pleasure it gives rise to). In judgments of cognition imagination does not have this kind of freedom and does not make this kind of contribution, neither in empirical nor in pure and mathematical matters. Unfortunately, for Kant mathematics drops completely out of the picture in this new, aesthetic context of the third *Critique*. Mathematics cannot be beautiful, Kant claims, contrary to what he had said in his earlier writings (before the mid 1780 s, see Wenzel 2001). Yet Kant does not view this as unfortunate. He thought so highly of mathematics that he wanted to save it from fashionable and unstable ways of thinking and mere talking. For him, mathematics and the mathematical sciences were *fortunately* saved from wild speculations and from people who do not know what they are talking about (Wenzel 2001). In this Kant had a point. But I think he could have instead, or in addition, used his theory of free play between imagination and understanding to find out more about how mathematics is actually done and what its nature actually is. His insights into the loose interplay between imagination and understanding and the aesthetic playfulness of their interaction, when each plays, so to speak, with the function of the other, are after all rich insights.

In doing mathematics we often go by examples. We use them as models and do so in analogical ways when looking at new objects, problems, and open questions. Under the influence of John Locke and Alexander Baumgarten, Kant was aware of this aspect (Koriako, pp. 156–61). But he could have developed it further than he actually did. We are often dealing with particular instances, looking for suitable rules, and then use reflective judgment (*reflektierende Urteilskraft*), i.e., the power of judgment in its reflective function. The object at hand might call for new rules and a new framework. Familiarity with rules is necessary but not always determining. Although we often do mathematics mechanically and "blindly", as Wittgenstein said, we do not always do it in this way. We are never completely blind. Learning is more than drill (*Abrichtung*). Doing mathematics is more than carrying out mechanical calculations to get the bill right. We sometimes play with rules and instances. We try out new things. Learning mathematics and doing research are done in ways that go beyond mechanisms and rule-determined behavior, and I think this does matter to the development of mathematics and therefore in the end to mathematics itself (what mathematics we end up having). Children and researchers make use of playful features when speculating about the applicability and suitability of rules. This is more than mechanical and blind trial and error. We use examples we already know and we use them in analogical and ingenious ways.

In doing mathematics we are neither *blind* as Wittgenstein portrays it, nor are we *determined* as Kant thinks. Mathematics is not as fixed as Kant tends to think it is. Had he been more aware of the factors of interest, change, indeterminacy, and creativity, I think his theory of free play in aesthetic judgments would have allowed him to make interesting contributions to the nature of mathematics.

Imagine you work with the real numbers and think about an equation that has no solution in this number field, such as $x^2+1=0$. You think about somehow extending this field in order to have solutions for the equation at hand. You start with the real numbers and imagine ways of introducing new numbers. In contemporary mathematics we have the complex numbers, a number field that is an extension of the real numbers. There are different ways of introducing these numbers, relying on new notations (when a complex number is represented as an ordered pair of real numbers), or relying on geometry (when a complex number is represented as a point, or vector, in a two dimensional plane). The resulting structures are isomorphic, and happily in this new field every polynomial equation has a solution, which is what we were looking for. The complex numbers are what we call "algebraically closed". But something had to be given up. The real numbers are "ordered". They can be represented along a straight line, with negative numbers on the left and positive numbers on the right, such that the ordering respects addition and multiplication (sums and products of positive numbers are again positive numbers, that is, if $x<y$ and $y<z$, then $x<z$, and if $x<y$ and $z>0$, then $xz<yz$; see Lang, p. 390). But no such ordering is possible for the complex numbers. Thus if we originally assumed that ordering was essential for numbers, we now either have to say that the complex numbers are not numbers any more, or we have to drop this assumption. Mathematicians have chosen the latter way. The complex numbers are now considered to be "numbers", not the real numbers, but still really numbers. The axioms we now have for number fields usually do not include well-ordering. We cannot have everything. If we ask for a further extension, beyond the complex numbers, further assumptions will have to be given up. The so-called "quaternions" are an extension of the complex numbers and they can be realized in four-dimensional space over the real numbers, but now the axiom of commutativity has to be given up. For quaternions it is no longer true in general that $ab=ba$. Should we then still call them "numbers"? Well, this is up to us again. Of course one might say that this does not matter to mathematics itself, that this is all only about names and words. But I am trying to show that it matters whether we use *this* mathematics (writing textbooks focusing on the complex numbers) or *that* one (writing text books focusing on the real numbers, or the quaternions). It matters what we use. It matters regarding where we go and where we end up.

When learning or doing research, on the one hand imagination plays with various possible functions of the understanding when asking for suitable concepts and rules. On the other hand understanding guides our way of looking at the case at hand and demands from imagination to fill in and to create what the rules require. Methods of trial and error are used and often pleasure arises. I think more than pleasure in success is involved here.[5] I think the pleasure involved in doing mathematics can also be *pleasure in the play itself.* Even if playful attempts or acts of contemplation do not yield the desired result, we still

---

5 The pre-critical Kant could still say "demonstrations in geometry can be beautiful [*Schönheit haben*] due to their shortness [*Kürze*], their completeness [*Vollständigkeit*], their natural light [*natürliches Licht*], and their suitability [*leichte Faßlichkeit*] for an easier understanding" (*Anthropology Lecture Collins* 1772/73, AA 25: 177; translation mine, see Wenzel 2001, p. 417). The Kant of the *Third Critique* does not think like this any more. "Shortness" and "completeness" now are too objective. Ideas of "natural light" and "suitability" drop out. Maybe "natural light" is too poetic and "suitability" too much in the spirit of transcendental realism (which assumes that mathematics is out there, independently of our minds).

can feel the pleasure of adventure in trying out new things, new objects and new methods of learning that lead to new objects. The search itself has a value and I think the aesthetic aspects involved also matter to mathematics itself, because, again, they influence how and where we look and thus they matter to the question of what we find, accept, and choose to use – and thus now have.

One might still argue that such pleasure is not part of mathematics proper and that when playing with rules and instances one is not strictly speaking "doing mathematics" any more. In this view, only when it comes to the proofs, in the end, is one doing mathematics again. But I would counter such a view by pointing out that the results of playful considerations are often the new objects and that therefore the play itself matters. It literally does *matter*, because it leads to new objects, i.e. the subject matter for mathematical understanding. New methods can be agreed upon and then become part of the growing corpus of mathematical knowledge. Besides political, economic, geographic, and many other influences, there are also aesthetic influences that matter in the development of mathematics. They matter not as necessary means but as decisive factors.

Kant had an aesthetic theory according to which judgments of taste can be pure and *a priori*. Purity consists in disinterestedness regarding the existence of the object. We do not depend on the real existence of the object as we find it in the case of satisfaction for the agreeable (when the object affects us and our senses and sensory inclinations are involved), nor do we depend on it as is the case regarding satisfaction for the good (when a morally good action has to be produced or when something empirical is instrumentally good for something else). This freedom of interest, which Kant emphasizes in the first moment of taste, does actually suit mathematical objects. We do not depend on their existence regarding our sensation, and we do not need to bring them into existence. They are freely available for everyone. Furthermore, the *a priori* justifying ground, which Kant had found to underlie aesthetic judgments (subjective purposiveness), does suit mathematics, because mathematical objects can show new and unexpected features. One might argue that the purposiveness we meet with here tends to be objective purposiveness (between objects) and not subjective purposiveness (between the object and our subjective state of mind). But as I have shown above, it seems to me that the latter is involved as well.

In everyday life, imagination is necessary when we recall an object that is absent, when we recall a melody, or the face of a friend. Both reproductive and productive functions are then essential, as Kant has pointed out in the A-version of the Deduction of the categories. In the 20th century, studies of perception have shown that these functions of the imagination are also involved in real perception, when the object is actually present and we are not just imagining or remembering things. Also in real perception it is the case that we do not actually see everything we think we see. We literally think more than we see: We project and fill in a good deal in more or less conceptual ways. We rely on memory and expectation, invention and imagination. We take much for granted also in perception of real objects, and we do so by using imagination. Kant had a good idea of this when he saw in the imagination "a blind though indispensable function of the soul" (*Critique of Pure Reason*, section 10, A78/B104). But he did not develop this idea in his views on mathematics.

Some of Kant's observations about imagination do apply to mathematics as well, because also – and particularly – in mathematics we can ask how "absent" or "present" an object is. Imagination is productive (*exhibition originaria*) with respect to intuition in time and space, *a priori* and empirically, when we draw a line in our mind or on a piece of paper. Imagination is necessary to give us pure intuitions (*Anschauungen*) in mathematics. In order to think of a line or a triangle, we need to imagine them as drawn and constructed and as given totalities. We give them to ourselves. Imagination is necessary in order to give substance to abstract mathematical reasoning. Intuitions have to be built up, manifolds have to be synthesized and held together. Only then do our concepts have objects. Only then do our thoughts have substance to which they can be applied. We think of, and also somehow "see", a complex number, a continuous but nowhere differentiable function, or a linear algebraic group that is locally eight-dimensional over the complex numbers. Or should we rather say that we do not see anything at all in such cases and that mathematics is, as Kant claims, "nothing but rules", without any need of intuition? But Kant also famously says that concepts without intuition are empty. Thus we should see something after all. So what is going on?

There is a transcendental function of the imagination to insure unity in apprehension, production, reproduction, and apperception. This basic and productive function is part of imagination in figurative synthesis, and we can turn to this function (*Critique of Pure Reason*, section 24) to explain that, when considering very abstract and higher-dimensional mathematical objects, we still "see" something and concepts are not empty. But this transcendental function by itself is not enough. There is also productive imagination, which generates methods (schemata) and images (A 140–41/ B 180) that are called upon in reproductive imagination. Productive and reproductive imagination are intertwined. It is not just all rules. Even chance gets involved, when we do things "spontaneously" in the sense of "*unwillkürlich*". Kant calls this fantasy (*Phantasie*). Imagination then is creative, "poetic" (*dichtend*), and playful: "*die Einbildungskraft (als Phantasie) spielt*". I think we can be creative and poetic like this while still staying within the realm of *pure* intuition and *pure* imagination. This activity does not need to be merely transcendental in the sense of being a necessary condition of experience. It goes beyond that, and it does so also in mathematics, because examples, chance, attempts, and analogies are involved, as I have explained above. The activity and its results can still be pure (non-empirical). Kant has argued that there are metaphysical foundations of the natural sciences, and I am here suggesting that there are foundations of mathematics that also involve choice, variation, playfulness, and beauty in mathematics.

Kant does not only have room for, but he also has theories of, pure intuition, *exhibitio originaria*, and *a priori* functions of the imagination. The latter don't need to be completely governed by the understanding. Imagination can be productive, and this plays also a role in our doing mathematics, when we use examples and reach out to intuitions that go beyond the rules and concepts we started out with. These examples and models sometimes do not perfectly fall under the relevant rules, and when this happens they will suggest something new. Chance is essential in this, and in doing research we often get new insights from *this very surplus*, from wrong applications, borderline cases, and *Einfälle* from somewhere else. We often hold on to concrete examples, schemes, and images, even if they do not perfectly fit the rules and only serve as substitutes. The outer then might become part of the inner.

Through productive imagination and the element of chance we give something to ourselves in intuition. We can thereby experience the surprise of fitting and the encounter of "lawfulness without law" also in non-empirical matters and subjects such as mathematics. We can generate something new, allowing for new choices and new chances, and we can freely recall previous results. This encounter is both active and passive. It is discovery and invention at the same time. It is a form of *self-affection*, as we find in the enjoyment of playful fantasy when surprise is always around the corner. Through productive imagination we give ourselves pure and "concrete" intuitions in time and space. Think of the Latin root: *concrescere*, to grow together! We make something pure "grow". It is not the result of processes of abstraction from empirical matters, but it is pure creation. Then we are affected by our own activity and the result of it. This goes beyond the rules of mathematics that we started out with and have used so far. We can produce something new, a surplus, which does not need to accord with those rules.

Besides imagination as it is involved in chance and exemplarity, there is another way in which intuition and imagination do matter. Now I turn to arithmetic. We use symbols in mathematical notation, and good notation is essential in doing mathematics. We need to hold on to something. Reference must be fixed in some way (Wenzel 2010). The question here again is how independent or dependent, and how essential or inessential, symbolic notation is regarding the rules of our understanding and of mathematics itself. Answering this question will tell us how free imagination can be while still contributing substantially to mathematics.

Kant's notions of figurative synthesis and of monograms involve mathematical and linguistic aspects (see Makkreel, pp. 31–33). In the use of mathematical notation we can see both of them at work. Here we find letters and symbols (the linguistic side) and the meanings they convey (the mathematical side). We manipulate the visible signs and use the result to refer to something in the abstract object again. Kant was aware of this already in 1764 (Makkreel, p. 34).

Kant was also aware of right-left symmetries and problems of conceptually indiscernible differences of non-identical entities (Wenzel 2010). But he thought he could solve these problems with his theory of intuition and mathematical construction. Besides rules, also intuition and imagination are needed in making distinctions. (Kant's way of arguing here again shows the influences from Locke and Baumgarten; see Koriako pp. 156, 166–77). But in all this, for Kant intuition and imagination function only subserviently. In his view, imagination is told by the understanding what to do. As an example of this, let us again look at a mathematical object that was discovered after Kant's time – namely, the complex numbers. Algebra tells us that if there are any square roots of -1, then there will have to be exactly two of them. Understanding then asks imagination (algebraically so to speak) to fix a referent as *one* square root of -1 and to fix *another* referent as the other square root of -1. We call, or denote them by the symbols, "i" and "$-i$". Through rules of the understanding (algebra) we know that there must be two distinct roots, and thanks to intuition and imagination we now actually "have" them. Here imagination and intuition seem to play only subservient roles. It seems to be the understanding that sets the course and that determines what mathematics is. In this way we can understand, retrospectively, why Kant had no room for beauty in mathematics. Imagination did not matter. It had no choice. Hence Kant had no room for a free play of imagination and understanding when dealing with mathematical objects *as such*. But contrary to Kant, besides the commonsensical views that mathematics can be beautiful and that there are geniuses also in mathematics, a view Kant knew and had previously embraced himself, we now know facts about the nature of mathematics (proper) that Kant did not know and that shake the foundations of his views of mathematics. Since Gödel's 1931 results we know that axiomatic systems can be incomplete. We know that rules do not determine everything. The situation has changed, because on top of the question of which rules or axioms to accept, we now have the additional problem that rules do not settle everything even within their own domains. The questions have become more complicated and pressing.

Understanding cannot any more be seen as having full control even in matters of mathematics. Intuition and imagination can thus play more than subservient roles, and once imagination's freedom and contribution to mathematics is secured, so is the possibility of a pleasurable and meaningful free play of the faculties that is the ground for beauty and that can lead, by chance, to results that are mathematically relevant. To the newcomer the traces might be lost. But there is always room for new discoveries.

One might still try to drive a wedge between aesthetics and mathematics, arguing that the free play is one thing and its mathematical result is another. But in doing mathematics we go to and fro. We shift from one to the other, between the free play and the result, the example and the rule, intuition and logic, geometry and analysis (compare Poincaré on geometrical and analytical minds, Poincaré pp. 27–40). The two are intertwined. The empirical leaves its traces in what is realized from what is *a priori* possible. As there are prototype-theories of concepts in philosophy and the cognitive sciences, showing that perception and pictures (images) matter for concepts, so I am suggesting that mathematics is not just rules but that it is done with examples in mind and that this plays a role regarding what mathematics we have. Intuition, imagination, and aesthetics have left their traces.

For comments I wish to thank Riccardo Manzotti, Gottfried Gabriel, Sebastian Gardner, Joel Schickel, Kurt Walter Zeidler, and Michael Thompson.

References

Carmichael, Robert Daniel (1930): *The Logic of Discovery*, Open Court Publishing Co..

Changeux, Jean-Pierre, and Alain Connes (1989): *Matière à pensée*, Èditions Odile Jacob.

Changeux, Jean-Pierre, and Paul Ricoeur (1998): *Ce qui nous fait penser. La nature et la règle*, Éditions Odile Jacob.

Crawford, Donald W. (1982) "Kant's Theory of Creative Imagination", in T. Cohen and P. Guyer (eds.), *Essays in Kant's Aesthetics*, University of Chicago Press, pp. 151–178.

Ekeland, Ivar (1984): *Le Calcul, L'Imprévu: Les figures du temps de Kepler à Thom*, Editions du Seuil.

Friedman, Michael (1992): *Kant and the Exact Sciences*, Harvard University Press.

Gabriel, Gottfried (1997): *Logik und Rhetorik der Erkenntnis: Zum Verständnis von Wissenschaftlicher und Ästhetischer Weltanschauung*, Schöningh.

Gabriel, Gottfried (2011): "Kreatives Denken. Über den 'Geist' in den Naturwissenschaften", in *Phantasie und Intuition in Philosophie und Wissenschaft*, edited by Gudrun Kühne-Bertram and Hans-Ulrich Lessing, Königshausen & Neumann, pp. 199–214.

Gibbons, Sarah L. (1994): *Kant's Theory of Imagination: Bridging Gaps in Judgement and Experience*, Oxford University Press.

Giordanetti, Piero(1995): "Das Verhältnis von Genie, Künstler und Wissenschaftler in der Kantischen Philosophie. Entwicklungsgeschichtliche Beobachtungen," *Kant-Studien* 86, pp. 406–30.

Iseli, Rebecca (2001): *Kants Philosophie der Mathematik: Rekonstruktion – Kritik – Verteidigung*, Berner Reihe Philosophischer Studien 27, Verlag Paul Haupt.

Kant, Immanuel (1998): *Critique of Pure Reason*, translated by Paul Guyer and Allen Wood, Cambridge University Press.

Kant, Immanuel (2000): *Critique of the Power of Judgment*, translated by Paul Guyer and Eric Matthews, Cambridge University Press.

Kitcher, Philip (1984): *The Nature of Mathematical Knowledge*, Oxford University Press.

Koriako, Darius (1999): *Kants Philosophie der Mathematik: Grundlagen, Voraussetzungen, Probleme*, Kant-Forschungen 11, Felix Meiner Verlag.

Lang, Serge (1984): *Algebra*, Addison-Wesley Publishing.

Makkreel, Rudolf (1999): *Imagination and Interpretation in Kant: The Hermeneutical Import of the 'Critique of Judgment'*, The University of Chicago Press.

Parsons, Charles (1980): "Mathematical Intuition", *Proceedings of the Aristotelian Society*, ns 80, pp. 145–68.

Parsons, Charles (1983): *Mathematics in Philosophy. Selected Essays.* Cornell University Press.

Parsons, Charles (1992): "The Transcendental Aesthetics", in Paul Guyer, *The Cambridge Companion to Kant*, Cambridge University Press, pp. 62–100.

Parsons, Charles (1998): "Intuition and the Abstract", in Marcelo Stamm (ed.), *Philosophie in synthetischer Hinsicht*, Klett-Cotta, pp. 155–187.

Poincaré, Henri (1970): *La valeur de la science* (first published in 1905), Flammarion.

Resnik, Michael (1997): *Mathematics as a Science of Patterns*, Oxford University Press.

Wenzel, Christian Helmut (2001): "Beauty, Genius, and Mathematics: Why Did Kant Change His Mind?" In: *History of Philosophy Quarterly* 18/4, October, pp. 415–32.

Wenzel, Christian Helmut (2005): "Spielen nach Kant die Kategorien schon bei der Wahrnehmung eine Rolle? Peter Rohs und John McDowell", *Kant-Studien* 96/4, pp. 407–426.

Wenzel, Christian Helmut (2010): "Frege, the Complex Numbers, and the Identity of Indiscernibles", In: *Logique et Analyse* volume 53, number 209, March, pp. 51–60.

Wenzel, Christian Helmut (2011): "Urteil". In: *Neues Handbuch Philosophischer Grundbegriffe*, P. Kolmer and A. Wildfeuer (eds.), Karl Alber Verlag, pp. 2284–2296.

Wenzel, Christian Helmut (2011): "On Wittgenstein on Certainty". In: *Epistemology: Contexts, Values, Disagreement*, Papers of the 34th International Wittgenstein Symposium, vol. 19, Kirchberg am Wechsel, Austrian Ludwig Wittgenstein Society , pp. 320–22.

Wenzel, Christian Helmut, "Mathematics and Aesthetics in Kantian Perspectives". In: *The Psychology of the Mathematician*, Stephen Krantz (ed.), Mathematical Association of America (to appear).

Gary Banham
The Transcendental Synthesis of Imagination

Kant discusses the notion of the transcendental synthesis of imagination at prominent parts of both the two editions of the transcendental deduction of the *Critique of Pure Reason*. Despite this, there is scant agreement concerning the importance of it in Kant's argument. There is not even a standard understanding in current secondary literature as to what the "synthesis of imagination" really amounts to. On the one hand, Martin Heidegger describes it in a way that makes it equivalent to temporality in general, whilst, on the other, Peter Strawson, at least at one point, dismissed it as essentially irrelevant to the understanding of the transcendental deduction.[1] Heidegger's conception of transcendental imagination is one that wrenches it out of the context of Kant's philosophical project and divorces it in particular from the understanding of the transcendental deduction. In the latter respect at least Heidegger's reading is of a piece with that of Strawson. In this piece I aim to revise and present in a shorter version the central argument of my book *Kant's Transcendental Imagination* in which I set out to provide a case for thinking of the transcendental synthesis of imagination as the centre-piece of Kant's transcendental deduction and as something the interpretation of which was requisite to understand the way Kant's transcendental psychology relates to his transcendental logic.[2] The point of this piece is to uncover the way that the description of transcendental psychology Kant provides gives a distinctive account of the genesis of experience

---

[1] For Heidegger's classic reading see Heidegger, Martin (1929): *Kant and the Problem of Metaphysics* (1990 trans. by Richard Taft). Indiana University Press: Bloomington and Indianapolis, and for an account of how Heidegger's reading is best read through *Being and Time* see Henrich, Dieter (1956): "On the Unity of Subjectivity" (1994 trans by G. Zoeller) in D. Henrich (1994): *The Unity of Reason: Essays on Kant's Philosophy*. Harvard University Press: Cambridge, Mass and London. For Peter Strawson's dismissive view of transcendental imagination and transcendental psychology generally see Strawson, P.F. (1966): *The Bounds of Sense: An Essay on Kant's Critique of Pure Reason*. Routledge: London and New York. Despite these claims in his most discussed work on the *Critique* Strawson later substantively revised his view of imagination.

[2] See Banham, Gary (2006): *Kant's Transcendental Imagination*. Palgrave Macmillan: London and New York, especially Chapters 1–4. This essay intends not only to provide a briefer version of the general argument provided in this book but also to show in a more basic way the reason for taking the argument I presented there as important for reassessing the significance of the transcendental synthesis of imagination.

that can be essentially expounded in a way that makes it analytically distinct from the argument concerning the categories.[3]

Kant's Symmetry Thesis

In order to begin motivating the discussion in question here I want to open with a citation that does not directly concern imagination but which raises a question that rather concerns the relationship between judgment and intuition. The reason for beginning with this citation is that the citation in question is one that certainly appears difficult to conjoin with another that I will later consider. The first citation arises within Kant's Metaphysical Deduction and it suggests an important connection:

> The same function which gives unity to the different representations *in a judgment* also gives unity to the mere synthesis of various representations *in an intuition*; and this unity in its most general expression, we entitle the pure concept of the understanding. The same understanding, through the same operations by which in concepts, by means of analytical unity, it produced the logical form of a judgment, also introduces a transcendental content into its representations, by means of the synthetic unity of the manifold in intuition in general. (A79/B105)

This statement gives what I will describe as *Kant's symmetry thesis*. The thesis claims that the unity of representations in a judgment and the unity of representations in an intuition is provided by what is termed "the same function". Despite making this claim in the first sentence of the citation, however, Kant also introduces a distinction in the second sentence, as, in the second sentence, he makes clear, that the unity of judgment is grounded on *logical form* whilst that of synthesis depends, by contrast, on a *transcendental content*. So the symmetry thesis is not one that claims that the "same function" works to produce the "same result" as the result is rather one that is different in the two cases. It is, nonetheless, surprising that the symmetry that is claimed here is one that can be said by Kant to exist.

---

3 The argument concerning the categories, I want to suggest, is one that could be built upon this genetic argument that arises from transcendental psychology. In making this claim I am arguing for something quite distinct from Heidegger's phenomenological reading since, as will become apparent, the relationship between transcendental imagination and transcendental apperception is central to my reading whereas Heidegger explicitly aims to sever the link between them.

What makes this claim a surprising one becomes clear when we put the citation from the Metaphysical Deduction alongside one apparently asserting something quite different in the B-Deduction. In this second citation we find a claim about intuition that does not reflect the symmetry asserted in our first citation. Kant writes here:

> Space, represented as *object* (as we are required to do in geometry), contains more than mere form of intuition; it also contains *combination* of the manifold, given according to the form of sensibility, in an *intuitive* representation, so that the *form of intuition* gives only the manifold, but the *formal intuition* gives unity of the representation. In the Aesthetic I have treated this unity as belonging merely to sensibility, simply in order to emphasize that it precedes any concepts, although, as a matter of fact, it presupposes a synthesis which does not belong to the senses but through which all concepts of space and time first become possible. For since by its means (in that the understanding determines the sensibility) space or time are first *given* as intuitions, the unity of this *a priori* intuition belongs to space and time, and not to the concept of the understanding (cf. § 24). (B160–1n)

In this second citation we find a claim concerning intuitive unity, which, whilst not directly contradicting the claim made in the first citation, certainly appears at variance with it. In this second citation we find that the unity of intuition discussed is one that makes all concepts of space and time possible, but, which itself, whilst presupposing a synthesis, a synthesis that involves understanding, does not depend on "the concept of the understanding". Since this intuitive unity is explicitly here said to *not* depend on the concept of the understanding it would seem that the unity in question, a unity produced by intuition, is something quite distinct from the unity of judgment and it is unclear how there could be a sameness of function between this intuitive unity and that of judgment. Secondly, the first citation described the sameness of function by means of the pure concept of understanding that is explicitly repudiated here as a source for intuitive unity. So there does appear here to be a prima facie clash between two positions, both of which are stated in the second edition of the *Critique*. It is to present a case for thinking of the transcendental synthesis of imagination as that which provides us with the means to resolve this prima facie clash that I will utilise the resources of a genetic transcendental psychology to show.

Kantian Intuition

In order to begin making a response to this problem it will be necessary to clarify three matters: the understanding of intuition, the account of judgment and, above all, so I will argue, the transcendental synthesis of imagination. Only as a result of taking each of these in turn will it be possible to arrive at a view of what the passage from the Metaphysical Deduction described as the "same function" at work in giving unity to both judgment and intuition. Pure intuition is, however, characterized in two different ways by Kant. At the opening of the Transcendental Aesthetic the criteria by which intuition is determined is that of immediacy since Kant here writes that intuition is in an immediate relation to that which is cognised and that thought, by contrast, is only indirectly related to the "matter" of cognition. (A19/B33). The contrast between intuition and concept is made explicit when the latter is described later as always in a mediated relation to objects (A68/B93).

If the first criterion of intuition is one of immediacy, a subsequent account of intuition emerges that stresses singularity instead. This second criteria emerges at the conclusion of Kant's arguments with regard to the status of space and time as *a priori* intuitions, where their intuitive, rather than conceptual character, is stressed by means of it. When providing this argument Kant writes the following with regard to space: "every concept must be thought as a representation which is contained in an infinite number of different possible representations (as their common character), and which therefore contains these *under* itself; but no concept, as such, can be thought as containing an infinite number of representations *within* itself. It is in this latter way, however, that space is thought" (B40 and compare B48 for the parallel treatment of time). The singularity of the intuitive representation of space and time precludes the class-individual relationship at work for concepts as all the parts of an intuitive representation belong within a singular givenness of this representation.

These two distinct criteria of intuition have produced very different accounts of the logical sense of it. Stress on the singularity of intuition has tended to assimilate it to a certain type of concept, namely demonstrative terms and in quasi-Hegelian fashion, to the simple ability to refer to *thises*.[4] Others, by contrast, have favoured Kant's initial criteria of immediacy in terms of a transcendental regressive argument.[5] If the problem of the former reading is that it appears to

---

4 The Hegelian turn is explicit in the work of Sellars, Wilfrid (1968): *Science and Metaphysics: Variations on Kantian Theme.s* Routledge & Kegan Paul: London and New York, and for a logical rendition of intuition that takes off from the singularity criteria of it see Hintikka, Jaako (1969): "On Kant's Notion of Intuition (*Anschauung*)". In: T. Penelhum and J.J. MacIntosh (eds.) *The First Critique: Reflections on Kant's Critique of Pure Reason.* Wadsworth Publishing Company: Belmont.

5 For examples of accounts of intuition that are based primarily on Kant's immediacy criteria see Falkenstein, Lorne (1991): "Kant's Account of Intuition", In: *Canadian Journal of Philosophy* 21:2 and McDowell, John (1998): "Having the World in View: Sellars, Kant and Intentionality". In: *The Journal of Philosophy*, XCV: 9.

undercut the heterogeneity between intuition and concept that is stressed repeatedly by Kant, the problem with the latter is that the immediacy invoked with regard to intuition is one that involves formal principles in some sense.[6] Neither view appears sufficient, and yet it is also unclear how to combine these two criteria which appear, at least on standard readings, very different. This suggests that there is something about these readings of Kantian intuition that is faulty and that a different characterization of it is required. It is not, however, by attending to the criteria of intuition apparently provided in the argument of the Aesthetic that such a different account can be set out. I will return to indicating a different way of viewing intuition below when I arrive at the basic characterisation of imagination.

Kantian Judgment

Kant's Metaphysical Deduction appears to offer some kind of argument for deriving the categories of pure understanding from the table of judgments. It is, to say the least somewhat controversial as to whether Kant succeeds in this task.[7] However, whilst it would require an extended discussion to investigate this question, what is apparent is that the A-Deduction, at least, involves scant reference to the understanding of judgment broached in the Metaphysical Deduction. The B-Deduction, by contrast, contains, in §19, a discussion of objective judgment that is clearly linked to at least the opening strategy of the deduction.[8] However, regardless of the question of whether the B-Deduction is best approached through a division of "steps", the role of the reference to judgment in §19 is controversial in at least one important way, namely, by the suggestion that the discussion of

---

6 This latter point is particularly stressed in Caygill, Howard (1995): *A Kant Dictionary*. Basil Blackwell: Oxford, pp. 265 – 6.

7 Until recently the vast majority of commentators took the argument of the Metaphysical Deduction to be a failure and even a somewhat dismal one. Representative here would be the view offered in Bennett, Jonathan (1966): *Kant's Analytic*. Cambridge University Press, Chapter 6. There was, however, an older tradition of appreciation of its argument as represented by Reich, Klaus (1932) :*The Completeness of Kant's Table of Judgments*. (1992 trans. by J. Keller and M. Losonsky). Stanford University Press: Stanford. Interest in its argument has also been stimulated more recently by the work of Longuenesse, B. (1993): *Kant and the Capacity to Judge* (1998 trans. by C.T. Wolfe) Princeton University Press: Princeton. Longuenesse's rich work would deserve an extended response which I hope to undertake elsewhere.

8 The internal structure of the B-Deduction has conventionally been described as containing "two steps" at least since the article by Henrich, Dieter (1969): "The Proof-Structure of Kant's Transcendental Deduction" In: *Review of Metaphysics* 22: pp. 640 – 59.

judgment within it is not congruent with the division of types of judgment Kant provides in the *Prolegomena*. Describing the relationship between these two accounts of judgment will help us to arrive at a view of the place judgment has within Kant's conception of the B-Deduction and will enable us subsequently to relate this conception to that of transcendental imagination.

In §18 – 20 of the *Prolegomena* Kant makes a distinction between two types of judgment, judgments of perception and judgments of experience. Judgments of experience are described as being possessed of "objective validity" and as involving the categories. By contrast, judgments of perception involve only "the logical connection of perception" and not the categories strictly speaking. Not only is it the case that Kant makes this distinction but he also goes on, in §18, to provide a genetic account of how judgments of experience are arrived at as a consequence of our first making judgments of perception. So the argument proceeds from the basis that there are types of judgments made that do not require the categories and that these judgments are, in fact, genetically the primary judgments that are made in relation to affective phenomena. Finally, a last point is made that the argument goes on to utilise. This is to the effect that some judgments of perception are intrinsically incapable of becoming judgments of experience whilst others, by contrast, are capable of being developed so that they become judgments of experience.

In looking at the way judgments of perception are treated we will find that there is much to connect them to the way the imagination functions. So the first set of examples of judgments of perception concerns the following claims: "The room is warm, sugar sweet, and wormwood nasty", claims that are not expected to always hold universally even for the one making them let alone to bind others to agreement. In each case, says Kant, we only have "a reference of sensations to the same subject" and that only in "my present state of perception" (Ak. 4: 299). So there are two elements to these judgments that are isolated as occurring within them. The first is that they concern only the *matter* of intuition, sensation, a matter that is experienced affectively according to empirically subjective conditions. The second is that the relation of sensations to those experiencing them is one that has differential ways of being given at distinct moments of experience. So the judgment with regard to these sensations is one that is not only *not inter-*subjective, it is also not one that holds intra-temporally.

When Kant goes on to compare the judgments of perception given in these examples with judgments of experience he indicates that the judgments of perception concern only a *comparison* of perceptions with each other in a certain consciousness of one's state. By contrast, a judgment of experience is said to involve something quite distinct, namely, a relationship in which the phenomena is related to what is here termed "consciousness in general". The point of the contrast between these types of judgment is to make a critical reflection upon the way that experience is accounted for within certain (unspecified) other philosophies as is clear when Kant writes that: "it is not, as is commonly imagined, enough for experience to compare perceptions and connect them in consciousness through judgment" as from this alone there cannot be derived judgments that possess characteristics of universal validity or necessity.

For the latter to arise there has to be a relationship of the elements of intuition to the categories. In order to illustrate the difference Kant contrasts the judgment of perception that states that when the sun shines on a stone, it grows warm, with the judgment of experience that rather states that the sunshine *causes* the stone to grow warm by heating the stone. Here we see a type of judgment of perception that can become a judgment of experience. Notably, the difference between this judgment of perception and the earlier examples is that the warmth of the stone is neither described in terms of my affective state being one that is in question here and nor is the judgment merely restricted in temporal range. It is true that the sun does not constantly shine on the stone so there are temporal conditions of the judgment being true but these conditions are not ones that are true in terms of mere empirical subjectivity. Rather it is that the duration of the sun's shining on the stone determines exclusively the question of the stone's warmth. However the judgment of perception does not include the sense of the necessity that is meant to apply to the observation of the warmth of the stone because it lacks reference to the pure concept of causation.

Now the distinction between judgments of perception and judgments of experience clearly indicates that Kant takes the view, in the *Prolegomena*, that there are judgments that can be and are made in experience that do not require the categories and that such judgments are merely empirically subjective. However, it is often thought that this view is at variance with the claim made in the *Critique*, principally in the B-Deduction. So, for example, Paul Guyer writes concerning the relationship between the two views: "The distinction is not to be reconciled with a deduction based on the categorial basis of all apperception, but only superseded by it".[9] This claim, in the form at least that Guyer makes it here, turns on a view as to how the transcendental unity of apperception relates to the conditions of judgment and I don't wish yet to investigate this but shall return to it subsequently. Prior to doing so I wish next to look at the basic description Kant gives of imagination.

---

9 Guyer, Paul (1987): *Kant and the Claims of Knowledge.* Cambridge University Press: Cambridge and New York, p. 101.

Imagination, Presence and Quantity

Kant describes the activity of imagination at B151 when he terms it the faculty of "representing in intuition an object that is *not itself present*". The ability to be able to represent something that *is* not present has to be understood in two respects. On the one hand, what is represented is not "present" in the sense of not being before one, that is, it is not there in front of you or anywhere near you. This is equivalent to saying that there is no "space" that is occupied by that which appears to be represented and thus that the representation is one that is distinct from that which we understand to be the "normal" means of presentation that is sensory. If this is one of the aspects in which the "object" in question is not "present", however, it is also necessary to think of that which imagination represents to us as something that is not occurring now, that is, as not something given to us *in the present*. Thus what is represented is either something that *was* or something that *may come to be* rather than being there right now. If we hold together these two respects of non-present then the "object" of imagination is either temporally or spatially removed from us and could be both of these despite being, in some way, "represented".

This basic characterisation of the "power" of imagination appears, once it is unpacked, to be more complex than we might initially suspect since the ability to be able to represent to ourselves something that *is* not invokes both the elements of intuition and in so doing shows that whatever "representation" is, it is something that is not merely privative in relation to intuition but also formative of it in some way. Without the ability to relate to that which *is* not there would be a problem with identifying anything as being-there. However, I wish, prior to undertaking the kind of investigation that will enable a more detailed view of the role of transcendental imagination, to first lay out a part of the power of imagination that is rarely attended to but which is specifically and explicitly pointed to by Kant. This concerns the presentation of "images".

If the description of the activity of imagination at B151 led us towards some interesting initial points I want now to connect this to how Kant describes imagination at A120 where he determines it as the "active faculty" for the synthesis of the manifold. After so describing it Kant gives the following characterisation of the relationship between imagination and an image: "Since imagination has to bring the manifold of intuition into the form of an image, it must previously have taken the impressions up into its activity" (A121). If we leave aside for now the question of what the "previous" action Kant refers to here involved, what we can see here, is that the result of this activity appears to be the formation of an "image" of the manifold. Kant subsequently describes the formation of this "image" as the arrival at time and space as ways of presenting magnitudes and objects of the senses (A142/B182). Such a formation of images is sharply distinguished by Kant from how a schema operates though the way this distinction is drawn is itself less than obvious. If the "image" of all magnitudes is space, the "schema" of magnitude, says Kant, is number. Concerning the "schema" of magnitude Kant writes the following:

> Number is therefore simply the unity of the synthesis of the manifold of a homogeneous intuition in general, a unity due to my generating time itself in the apprehension of the intuition (A143/B182).

The argument that "number" is the unity of the synthesis of the manifold involves a claim concerning the relationship between identity, iterability and unity that is central to the understanding of Kant's overall claim about the nature of experience. To put the matter in a nutshell, experience is grounded on synthesis and synthesis is, first and foremost, a quantitative achievement that works through the ability of imagination to present something that *is* not. So the achievement described in the account Kant gives of number is one that is at the heart of his claim concerning the possibility of experience itself. Putting the matter differently, in order to begin to lay out the exposition, it will be my claim that the ability to have "representations", that is, the ability to present something regardless of whether the something in question is either nearby or taking place now, is the same ability as that which makes experience itself possible. This ability is the one that is at issue in being able to identify and repeat the unitary recognition that states that something in particular, some particular and not another, is before one. And this ability is, likewise, that which enables us to say that it is legitimate for us to use categories in experience. Once put like this, the argument appears deceptively simple. As I will argue in some detail now it is not simple but the heart of the claim is one that requires an understanding of imagination as that which is "the pure form of all possible knowledge" (A118).

Particulars, Identification and Time

In the A-Deduction Kant makes two opening moves, the significance of which I wish to begin by pointing to, in order subsequently to connect these moves to some parallel passages in the B-Deduction. Kant opens the preliminary version of the A-Deduction by pointing to how our cognition is subject to time as in time all must be "ordered, connected, and brought into relation", a point that, he claims, "must be borne in mind as being quite fundamental" (A99). Having opened with this point about temporality Kant proceeds to first of all discuss the notion of how we can "represent" a manifold at all writing the following significant sentence:

> Every intuition contains in itself a manifold which can be represented as a manifold only in so far as the mind distinguishes the time in the sequence of one impression upon another; for each representation, *in so far as it is contained in a single moment*, can never be anything but absolute unity. (A99)

Immediately after this sentence Kant describes synthesis as involving running through and holding together so that a single manifold can be represented *as* a manifold. The presentation of a manifold *as* a manifold requires such a synthetic act and thus the condition of individuation of intuitive representations is that such synthesis takes place.

This opening claim of the A-Deduction can itself be re-presented if we assess what has been said in light of my previous suggestions. The basic imaginative act is being able to "represent" that which *is* not. In a generic understanding representation thus simply is the sense that what is able to be given to one is not equivalent to the presence of anything directly before one. So representation generically is produced by imagination. This generic claim about representation is illustrated in the opening move of the A-Deduction with regard to how it is that, intuitively, it is possible to distinguish elements from each other. If there is an intuitive unity, if, that is, there is some particular given to one, then this requires that such a particular can be distinguished from another one and Kant's claim at the opening of the A-Deduction is that such distinction has itself a given condition, namely, that the time "in the sequence of one impression" can be made separate from that of a subsequent impression. Without this the ability to discern the way different impressions are sensory data of distinct sorts could not function.

So the basic claim of the opening move of the A-Deduction is that intuitive identification of particulars involves separation of one particular from another and that such identification is itself based upon the grasp that there are discrete temporal moments of givenness of any particular, a grasp that is itself the basis not merely for the identification of this particular *as* particular but also for the differentiation of it from other particulars. This is an evidently ambitious philosophical claim with which to open the A-Deduction but it also immediately shows a connection of the A-Deduction to my opening observations concerning imagination when we note that unitary awareness of a particular intuitive manifold indicates that an image has been formed *of* such a manifold in terms of its quantity, and, that there is first therefore a sense of temporality given as primary data, even for the possibility of the basic sense of a particular intuition to arise.

This is clearly part of what Kant meant by preparing for this initial claim by stating that cognition is subject to time and that within it all "must be ordered, connected, and brought into relation". Imaginative representation is of that which *is* not. Such a sense of what *is* not refers us to the view that what *was*, now no longer is. However to relate to that which *was* as something that now *is* not requires stabilisation of the flow of time sufficiently for individuals to be represented within it.

For various reasons Kant alters the mode of presentation of his argument between the A-Deduction and the B-Deduction but rather than attempt here to discuss the ways in which this makes the arguments distinct from each other what I wish to bring out instead is how motifs treated in one are repeated, if formulated at different points, in the other. The claim we have looked at here with regard to the opening of the A-Deduction has an important parallel towards the conclusion of the B-Deduction. In §26 of the latter Kant discusses the possibility of noticing the connection between two states of water, one where it is fluid and a second where it is solid due to having frozen. In examining this example Kant writes:

> in time, which I place at the basis of the appearance [in so far] as [it is] inner *intuition*, I necessarily represent to myself synthetic *unity* of the manifold, without which that relation of time could not be given in an intuition as being *determined* in respect of time-sequence (B162–3).

Here we find a repetition of the point made at A99 albeit in a distinct register. Just as at A99 we found that a single moment gives only a sense of "absolute unity" (inasmuch as it requires for a sense of a given intuition to emerge within it a synthetic act) so here we look at how the relation between one moment and another is at the root of the sense of change. The differentiation between the distinct states of the water requires the sense that the manifold that is given to us as the intuitive sense of the water be united in one way as occurring at one moment and in another way occurring at a distinct moment. Such a view of a relationship between distinct particulars as grounded on the temporal sense of their appearance is precisely the point of A99. Just as we saw in the analysis of A99 that what this requires is that imagination is at work in the enduring sense of individuation, so here we also note that it is also required for the view that there is a relationship between that which was (hence *is* not) and that which now is presented.

Now to build the argument on from this stage requires attention to something that is picked out by Dieter Henrich as "the problematic engendered by the concept of an object" and which is, according to Henrich, evident to Kant by means of an "elementary assumption" that Henrich describes as follows: "the primary occurrences of the real for cognition are presentations of simple qualities".[10] The reference to how a single moment provides and can provide only "absolute unity" that Kant makes at A99 is not elaborated on at this point of the transcendental deduction. However, in bringing out what is meant by this claim we can begin to develop the first suggestion we have uncovered here of how imaginative synthesis is part of Kant's general picture of the genesis of experience. In doing so, however, we will need to move away from the transcendental deduction, a move that I will justify in terms of connection of the opening suggestion of A99 to subsequent developments in the *Critique*.

Sensation and Continuity

The understanding of there being unity provided within an intuition inasmuch as it occupies only a moment is amplified when Kant discusses the Anticipations of Perception. Here Kant states that: "Apprehension by means merely of sensation occupies only an instant, if, that is, I do not take into account the succession of different sensations" (A167/B209). A singular relationship to a specific affection is given in the instant and this singular relationship, if taken alone, would be insufficient to develop a sense of the condition of identification of what was being taken to affect one. Kant states that if we isolate sensation in our analysis of experience it would be that which "does not involve a successive synthesis proceeding from parts to the whole representation" (A167/B209). So it would appear that analytic attention to sensation alone produces something like the "simple qualities" that Henrich refers to. Such simplicity is what is involved in Kant stating that sensations have an "intensive" rather than an "extensive" magnitude. How, though, is it possible to speak of them having a "magnitude" at all? Kant's response to this question is revealing:

> The absence of sensation at that instant would involve the representation of the instant as empty = 0. Now what corresponds in empirical intuition to sensation is reality (*realitas phaenomenon*); what corresponds to its absence is negation = 0. Every sensation, however, is capable of diminution, so that it can decrease and gradually vanish. Between reality in

---

10 Henrich (1976) p. 130 and Henrich also describes this assumption as one of "data-sensualism", a notion that would require its own analysis, not least to ask for whether, even should it be accurate, it would really be, as Henrich suggests an assumption that Kant so evidently shares as Henrich thinks with "the theory of knowledge of his time". Critical response to this suggestion would, however, require an extended comparison of Kant with, amongst others, Locke and Leibniz.

the [field of] appearance and negation there is therefore a continuity of many possible intermediate sensations, the difference between any two of which is always smaller than the difference between the given sensation and zero or complete negation. In other words, the real in the [field of] appearance has always a magnitude. But since its apprehension by means of mere sensation takes place in an instant and not through successive synthesis of different sensations, and therefore does not proceed from the parts to the whole, the magnitude is to be met with only in the apprehension. The real has therefore magnitude, but not extensive magnitude. (A167–8/B209–10)

Kant here presents a form of regressive argument. The assumption that there is a simple apprehension of sensation is one that requires that this simple apprehension be related to the basic possibility of how sensation is noticeable at all. It would not be noticeable if there were no distinction possible between any two given sensations. But, even though we assume that sensation has a simple apprehension we must still, even under this condition, account for the evident possibility of noting that sensations can be distinguished from each other. Further, the assumption of simple apprehension does not prevent us from stating the difference between distinct sensations. Rather, it points to how this difference is manifested to us, which is by means of distinct sensations having affects upon us that we can describe as different. Our description of this difference is that distinct sensations have degrees of affect upon us, degrees, that is, of intensity. One sensation appears to us as more or less intense as another and this difference is one that we can understand as proceeding to a point where the sensation can disappear entirely. This degree of difference is one, however, that can again only be noted by means of the presence of "intermediate sensations" that fill in the position between the intensities of any particular sensational experience and some unspecified other experience. Another way of putting this point is that given sensations, whilst taken as individual, may be analytically described as "simples" but the given sensations are never alone as they are part of a spectrum of possibility of givenness and this spectrum is what Kant refers to here as "a continuity".

So if the isolation of sensation leads us to a notion of sensation as a "simple quality" the simplicity of it is not one whereby any given sensation can be described as a part that is less complex than other parts. All parts of sensation are equal in the sense of being the same kind of magnitude. It is due to this that there is continuity between sensations. In this respect sensations, or the "matter" of intuition, as Kant often terms them, are structurally akin to the forms of intuition. Space and time are also continuous quantities in the sense that they cannot be broken down into simpler parts from which they can be built up.

> Such magnitudes may also be called *flowing*, since the synthesis of productive imagination involved in their production is a progression in time, and the continuity of time is ordinarily designated by the term flowing or flowing away. (A170/B211–12)

Both the form and the matter of intuition are continuous quantities that are produced for us *as* quantities by the action of imagination, an action that was described at A99 as a running through and holding together. By means of this synthetic action we are able to describe our experience of sensation quantitatively even though sensation, taken by itself, is not extensive. The quantity by which we can determine sensation is one of degree, a degree that involves us in determination of sensation by means of not only the unitary event which is at work in a particular noticing of sensation but also by means of connection of this to the continuity of relations to sensation that gives us one basic sense of what is meant by "experience" in general.

Concepts of Reflection and Comparison

It is now possible to begin to return to the contrast between judgments of perception and judgments of experience though not yet in order to relate this distinction to the account of judgment in the B-Deduction. What we have uncovered from our preliminary description of imagination and its relationship to perception is that imagination is involved in the basic relationship to data of sense that enables us to represent that which *is* no longer and, also, to that which is *not yet*. In other words, time determination is something that requires imagination just as we should expect from the opening claims of A99. However, returning to the examples of judgments of perception, the basic forms of which were apparently incapable of being turned into judgments of experience, what is apparent about them is they both involve only fleeting states of perception. Due to this it is the case that "they are not intended to be valid of the object" (Ak. 4: 299). They do require imagination even to be experienced as capable of providing us with judgments of perception since without imagination the comparison that is at work in such judgments would not be possible. Similarly the degree of perception of them is revealed to us in terms of the *quality* of the judgments in question.

However if comparison is required for judgments of perception then what is involved in making such a judgment is something more than merely experiencing the instant of perception and relating this to the continuity of sensation. A comparison such as is stated in the judgment requires explicit use of concepts. Without yet looking at how concepts relate to the objective judgments described in the B-Deduction I wish first to look at some concepts that are not categories and which will help us to arrive at a sense of how there can be concepts of understanding that are not equivalent to the categories. In the "Appendix" to the Transcendental Analytic Kant discusses "reflection" as an act that concerns "the subjective conditions under which [alone] we are able to arrive at concepts" (A261/B316). By following this root Kant distinguishes between some elements of logical form of concepts and the means by which the content of concepts differs when employed phenomenally from that which would apply to considerations of things-in-themselves. However, whilst this is the point of Kant's discussion the formulation of concepts that are of understanding but not categories is also stated to be part of the account he is giving as when he states that the "four headings of comparison and distinction" requires only "the comparison of the representations which is prior to the concept of things" (A269/B325).

Now, I want to suggest here that the concepts of comparison that are listed by Kant in the Amphiboly are ones that are not just related to the logical form of judgments but also state primary conditions of judgment. In making this claim I am following the explicit example of Salomon Maimon who wrote: "The forms of concepts in general are identity (unity in the manifold), but also difference, by means of which the manifold is thought as a manifold" and also stated:

> Difference and identity are the forms of perceptions in general (of individual sensible intuitions). When a perception, for example, *red*, is given to me, I do not yet have any consciousness of it; when another, for example, *green*, is given to me, I do not yet have any consciousness of it in itself either. But if I relate them to one another (by means of the unity of difference), then I notice that red is different from green, and so I attain consciousness of each of the perceptions in itself.[11]

Here Maimon repeats the point we derived above from our examination of imagination but inserts into the point the way in which the comparison that is made within a judgment of perception is possible. It is possible by use of the basic comparative concepts of identity and difference. It is through and by means of them that the judgment of perception is possible. So when we pass from the imaginative act of comprehending that there is a sensation that has a temporal condition to comparison of this particular sensation with another we have to utilise the concepts of identity and difference and it is these concepts, concepts of

---

11 Maimon, Salomon (1790): *Essay on Transcendental Philosophy* (2010 trans. by N. Midgley et. al). Continuum: London and New York, p. 74. The relationship between Maimon's *Essay* and the *Critique* would be worth extended attention on a future occasion. Beatrice Longuenesse's examination of concepts of reflection and comparison, which draws on an investigation of Kant's lectures on logic, makes a very similar point and for this see Longuenesse (1993).

understanding but not categories, that are decisive for the possibility of comparative comprehension of intuitions as particular individuals. The concepts of comparison become reflective when we connect them to consciousness in general and it is reflective awareness of the concepts that is at issue in the formation and use of categories in judgments of experience. The argument to this effect requires us to turn now to the description Kant gives in the B-Deduction of judgment and its relationship to the objective unity of apperception.

Judgment and Apperception in the B-Deduction

The first part of the B-Deduction contains an account of the transcendental unity of apperception and its relationship to judgment, as is recognised in standard readings of it. However, I want here to bring out firstly a question about how the transcendental unity of apperception is understood and secondly a way of connecting this understanding to a justification of the view that judgment as presented in §19, is described in way that does not conflict with the argument that there are forms of judgments that do not require the categories as is stated in the *Prolegomena* and as we have defended by use of the concepts of reflection.

In §16 Kant discusses what he terms "original apperception" and states that it is a self-consciousness which first *generates* the representation "I think" (B132) though he does not here state how this generation is itself possible. We will discover later that it requires a specific act of imagination for this to be possible but it is worth stating that it is here indicated that a generation does take place. The transcendental unity of apperception is connected here to the synthesis of intuitions in a way that mirrors a claim made in the A-Deduction. In the A-Deduction Kant wrote:

> the mind could never think its identity in the manifoldness of its representations, and indeed think this identity *a priori*, if it did not have before its eyes the identity of its act, whereby it subordinates all synthesis of apprehension...to a transcendental unity. (A108)

Here the apprehensive focus upon particulars is grounded in a transcendental unity that requires what appears to be a reflective sense of self-identity. This suggestion is presented in a parallel passage in §16:

> Only in so far...as I can unite a manifold of given representations in *one consciousness*, is it possible for me to represent to myself the *identity of the consciousness in* [*i.e. throughout*] *these representations*. (B133)

In following Maimon's lead we can see that the reflective conception of identity is at the root of the apparently self-reflective act that is engaged in when the transcendental unity of apperception is presented. The sense that there is a unitary awareness of the phenomena that is given is connected to the view that it is the same consciousness that is so aware. This relationship between the unity of that which is cognised and the unity of that which is cognising is at the heart of the claim that Kant makes for an "objective unity of apperception". Being aware of the claimed identity intra-temporally requires that there is the ability at any given time to bring the manifold to awareness as the unitary manifold that it is. We saw above that even the relationship to sensation is one that is based on quantity and continuity and this reference of the basic matter of intuition to conditions of a unitary sense of the form of intuition is here suggested, a suggestion that is later amplified in the second part of the argument of the B-Deduction, to be founded upon the working of a reflective sense of self-identity.

Now the problem with this claim is that it appears to require a view of self-identification to be given in some way separately or distinctly from the awareness of the manifold and yet to ground this awareness in the sense of the manifold. The latter point for example might be thought to be stated in the following: "Synthetic unity of the manifold of intuitions, as generated *a priori*, is...the ground of the identity of apperception itself" (B134). As we will see later this apparently opaque statement points to a view of the identity of apperception that helps us to understand its condition of being given as grounded in an imaginative act that is nonetheless only completed by the formation of the unity of apperception itself.

Leaving this point aside for now the point that Kant stresses in §18 is that the unity of apperception is that to which the manifold of intuition is related when we arrive at the concept of an object. Such a unity of apperception is distinct from any empirical notion of self-awareness, which latter involves associative connections. After making this point Kant turns, in §19, to his account of judgment. In doing so Kant stresses the difference between judgment and association stating that: "a judgment is nothing but the manner in which given modes of knowledge are brought to the objective unity of apperception" (B141). In making this claim about judgment Kant has appeared to many, including Guyer, as indicated in the citation from Guyer given earlier, to be suggesting that objective judgment is necessarily categorial given its connection to apperception.

Now what Kant discusses in the account of judgment that develops here is the way in which all types of judgment, regardless of their specific, often contingent content, have a necessary form. In a contingent judgment, such as the empirical claim that bodies possess heaviness, there is something necessary in their form that goes beyond the contingency of the content they are concerned with.

This resides in the way that such judgments depend upon the necessity of apperception in the synthesis of intuitions. Kant now contrasts this view of judgment with what takes place in associative combination when we merely state a connection that we do not suggest has to apply to the object given to us. Here what Kant describes as associative connection is much the same as what the *Prolegomena* termed a judgment of perception. However, whilst noting this suggests that judgments of perception are not true judgments on the model of the B-Deduction, I don't think that this is the right way to understand the claim that is being made in the B-Deduction. The claim that there is only an empirically subjective connection within a judgment is one that still requires that such a judgment possess the characteristics of logical universality and such a judgment also involves comparison. This indicates that even a judgment of perception has an objective character of a sort. Secondly, what judgments of perception do is raise the relationship to sensation that was discussed in the Anticipations of Perception to a reflective comparative sense so that we can state the view that the stone is warm as compared to its being previously cool. Such an awareness indicates a sense of the distinction between temporal states though it does not express a sense that the continuity of these states is one dependent upon a dynamical sense of connection.

In a basic way, then, I want to suggest that the associative connection discussed in §19 and paralleled in the account of judgments of perception, is one that not only requires a sense of the same logical form as is at work in judgments of experience but that the claim in the *Prolegomena* that there is a genesis of judgments of experience from judgments of perception is one that can be justified in terms of the text of the *Critique*. Before arriving at this point, however, it is worth adding that the claim that objective judgment has to be categorial due to its relationship to apperception is not as obvious as Guyer and others suggest it to be. The form of judgment that arises from combination of the manifold by apperception is one that can be expressed by and in a sense arises from utilisation of concepts of reflection and such concepts have a logical objectivity when they are used in a judgment to describe a state of affairs. The move from this to the view that the state of affairs being described is objectively part of the experience being undergone is what requires the categories or, as Kant puts in the *Prolegomena*, the judgment of experience.

Self-Affection, Imagination and the Unity of Apperception

The relationship between the unity of intuition and the unity of judgment is one that I now want to begin to approach by means of the account Kant gives of the connection between the transcendental synthesis of imagination and the transcendental unity of apperception. I mentioned above an apparent problem concerning how to understand unitary synthesis given that, on the one hand, it appears, on the basis of B133, that the unity of the manifold is dependent upon the identity of the consciousness in the synthetic representation of it whilst, on the other hand, at B134, Kant asserts the synthetic unity of intuition is the ground of the identity of apperception itself. How to square these apparent variant formulas given so quickly one after the other? In order to reach a response to this question, a response that will set us on the path of resolving the relationship between the passages I contrasted at the beginning of this piece, I will look now at Kant's key doctrine of "self-affection", a doctrine that takes us to the heart of the relationship between the transcendental synthesis of imagination and the transcendental unity of apperception.

In §24 Kant presents the basic picture of imagination that I re-traced above where he terms it the ability to represent in intuition something that is not itself present. The transcendental synthesis of imagination is also here described as "an action of the understanding on the sensibility" and as such an action it is "productive", not merely "reproductive". After making these points about imagination Kant next turns to resolving the "paradox" of how to account for our ability to represent to ourselves the manifold of inner sense. The reason that there is a "paradox" with regard to it is that "we intuit ourselves only as we are inwardly *affected*", which entails that we appear to have a passive relation to ourselves which yet is grounded in some sense on an active ability of representation. In explaining this paradox Kant unfurls his view of what the transcendental synthesis of imagination consists in and how it relates to the transcendental unity of apperception.

The relationship of understanding to inner sense is that the understanding has to take up the elements of intuition and bring them together. This requires synthesis and Kant here presents this synthesis in analytic isolation:

> Its synthesis....if the synthesis be viewed by itself alone, is nothing but the unity of the act, of which, as an act, it is conscious to itself, even without [the aid of] sensibility, but through which it is yet able to determine the sensibility. (B153)

The understanding has a basic capacity to present unity of determination in an act and to affect sensibility thereby. There are two important points to grasp concerning this, however, which is that the act in question is a self-affecting act and that such a self-affecting act is what is meant by the transcendental synthesis of imagination:

> the understanding, under the title of a *transcendental synthesis of imagination*, performs this act upon the *passive* subject, whose *faculty* it is, and we are therefore justified in saying that inner sense is affected thereby. (B153)

Kant also describes the transcendental synthesis of imagination as the "synthetic influence of the understanding upon inner sense" and the action of it is illustrated in §24 in terms of the ability to draw a line in thought and some other mathematical examples. However, more important than these examples are the claims that Kant makes here about "motion", where motion is presented not in terms of objects but rather as primary acts of the synthetic power and as productive thereby of "the concept of succession" as it is such "motion" that is involved in "determin[ing] the *inner* sense according to its form" (B155). The understanding produces the combination of inner sense, or, in other words, it provides us with the unity of temporality. Not only of temporality as Kant adds in a footnote:

> Motion....considered as the describing of a space, is a pure act of the successive synthesis of the manifold in outer intuition in general by means of the productive imagination, and belongs not only to geometry, but even to transcendental philosophy. (B155n)

The ability to describe a space, the ability, that is, to relate to the differential elements of space as parts of one space, is itself something that arises not merely from provision of the manifold of intuition but from an active relationship to this manifold such as is determined in synthesis. All of this appears to reinforce the message of B133 to the effect that the unity of the manifold is dependent upon the transcendental unity of apperception but not obviously to help with the alternate suggestion of B134 that the identity of apperception is itself, in some way, a consequence of the unity of synthesis. However, it is necessary to step back now and connect the description given here both to the general account of imagination and to the importance of the concepts of reflection. Since imagination enables a relationship to that which *is* not, the act of unification of the synthesis through use of this ability, requires the sense that not only what was *is* no longer, and yet can still be represented, but the view that this continuity of representation, despite the absence of what is represented, involves a distinction between what continues to be represented and other possible presentations and representations. It involves, that is, a sense of the relationship between "identity and difference", the primary concepts of reflection correctly picked out in Maimon's reading of Kant. So, for the transcendental synthesis of imagination to work, there has to be the ability to relate the primary concepts of reflection to the manifold that is given, and it is in and through the use of such concepts of reflection, that the identity of the act of apperception itself can take place. Hence, the right way to view the account of self-affection is as a process whereby the synthesizing subject relates itself to the positing of the elements of its synthesis by means of primary reflective concepts that lead to its becoming aware of itself *as* a synthesizer. Hence the identity of apperception is a result of the synthesis as is implied at B134 whilst also being that which, once the synthesis is complete, is the basis *of* the synthesis being unified. In order to clarify this point further it is now necessary to return to expounding the account given of the three-fold synthesis in the A-Deduction, an account that, I will suggest, is far from abandoned in the B-Deduction.

The 3-Fold Synthesis of Intuition in the Transcendental Deduction

Returning to the account of synthesis in the A-Deduction enables a fuller sense to be given to the transcendental synthesis of imagination than emerges just from §24 though it is notable that the argument of this section of the B-Deduction is to the effect that the transcendental synthesis of imagination is the basis for intuition being unified and that this occurs through an act of self-affection. The three-fold account of synthesis in the A-Deduction is presented in a confusing way in one respect, which is that Kant at each point of it describes the syntheses in question both empirically and transcendentally. Subsequently, when referring to some of them Kant speaks as if there is only an empirical form of the synthesis in question at issue. So reproductive imagination is described later as only empirical and apprehension is likewise so presented but these are short-hand ways of describing the empirical correlates of the transcendental syntheses.

The synthesis of apprehension at A99 was dealt with above and also shown to be amplified when Kant turns to describing the Anticipations of Perception. It is followed by an account of imaginative reproduction in which Kant speaks also about the ability to provide names to given events stably in terms of a rule "to which appearances are in themselves subject" (A101). This nominalization as related to the reproduction of appearances indicates that the relationship to stable appearances is one that is conceptually loaded. However when Kant describes the act of reproduction, what arises is an account that culminates in a description of the ability to represent unified representations of even the most basic forms of intuition:

> When I seek to draw a line in thought, or to think the time from one noon to another, or even to represent to myself some particular number, obviously the various manifold representations that are involved must be apprehended by me in thought one after the other. But if I were always to drop out of thought the preceding representations (the first parts of the line, the antecedent parts of the time period, or the units in the order represented), and did not reproduce them while advancing to those that follow, a complete representation would never be obtained: none of the above-mentioned thoughts, not even the purest and most elementary representations of space and time, could arise. (A102)

Thinking the time from one point to another or describing the space involved in the parts of the line are examples reached for again in the passage from §24 we have just been considering. Just as in the account from §24 we also here find it asserted that the "purest and most elementary representations of space and time" depend upon this synthetic act. The act that is here described is one of holding in thought "the preceding representations" as we move on to the succeeding ones. This act is identified here as the work of imagination, in accord with the basic characterisation of imagination we uncovered from B151. That which *is* not is retained and also held in continuity with that which comes forth next. Kant refers also to the ability to nominalize events as we describe what they are but such an ability requires conceptuality which is not itself an imaginative act. So what is required to be added next to the understanding of the genesis of experience is a further act of "re-cognition" which arises from and because of the imaginative ability to re-present that which has been given, thereby unifying the manifold.

The German term for "concept" is *Begriff*, a term which involves "grasping" something and Kant plays upon this when he describes the re-cognitive act that introduces conceptuality. Conceptuality is related to the ability not merely to reproduce the previous moment or to hold it together with the subsequent one but to be conscious that such acts are taking place. Such a consciousness is the basis of concepts: "the concept of the number is nothing but the consciousness of this unity of synthesis" (A103). The reason why number and other mathematical examples are given a primacy in Kant's account is because they explicitly involve re-cognition in their very presentation. A line has *to be drawn* and a number involves units of presentation explicitly. Further, the act of recognition that is at the root of apperception is one whereby the act of apperception is held to be a numerically identical one despite its intra-temporal character. The recognition of the act as a conceptually loaded one does not require that the transcendental unity of apperception itself involves categories even though the presentation of number itself does involve such. It is rather that it has to involve concepts of re-flection and that such concepts are primarily produced in the awareness of the imaginative re-collection of moments. So the imaginative act is one that we only hold analytically distinct from the transcendental unity of apperception when the two are effectively welded together.

The passages at B133 and B134 are correlates of equivalent passages at A118 and A119. At A118 Kant makes the point that the unity of apperception arises from the prior possibility of synthesis as suggested at B134 writing: "the principle of the necessary unity of pure (productive) synthesis of imagination, prior to ap-perception, is the ground of the possibility of all knowledge, especially of expe-rience". Here the unity of the synthesis of imagination is presented, analytically, as distinct from, and, in a sense, prior to the transcendental unity of appercep-tion. That the synthesis of imagination has to take place is indeed the self-affec-tive ground of the possibility of the transcendental unity of apperception. That this synthesis is unified is, however, not something that can be given prior to the arrival at the sense of the unity of apperception, a unity itself only possible through the awareness of concepts of reflection. The passage at A119 corrects the earlier formulation and in so doing arrives at a statement akin to that of B133:

> The unity of apperception in relation to the synthesis of imagination is the understanding; and this same unity, with reference to the transcendental synthesis of the imagination, the pure understanding.

In this statement the point is made that the pure understanding is essentially at one with the transcendental synthesis of imagination just as we discovered from B153. This point reinforces the view of B133 that the arrival at a unitary sense of the manifold only occurs by means of the transcendental unity of apperception, a view that itself is one we can assert however only once we have become aware of the working of the transcendental synthesis of imagination.

The three-fold synthesis that is presented at the beginning of the A-Deduc-tion is not repeated in this form at the beginning of the B-Deduction though Kant does state that combination includes "besides the concept of the manifold and of its synthesis, also the concept of the unity of the manifold" (B130), an in-dication that the three elements of the sense of the manifold, the synthesis of it and the unity of this synthesis are distinguishable and yet also have to be relat-ed. Similarly, at B132 Kant speaks of an original apperception as that which *gen-erates* the "I think", suggesting in this formula that the original apperception is one that is prior to the sense of the unity of it that is expressed in the "I think" and which, I am suggesting, would be a reflective sense of the basic concepts of identity and difference. As we have seen §24 introduces and explicates the transcendental synthesis of imagination as the self-affective ground of the possibility of the transcendental unity of apperception and the re-cognitive ability of the transcendental unity of apperception is related to judgment in §19, whose argument reprises the account of the synthesis of recognition that is given in the A-Deduction. The most elaborate description is performed of the synthesis of apprehension in §26 of the B-Deduction as here Kant moves between an account of empirical awareness and a crucial transcendental act. It is in the context of so doing that Kant presents the discussion of space and time at B160 – 1n that we found so difficult to relate to the symmetry thesis asserted in the Metaphysical Deduction.

In returning to the passage from B160 – 1n I wish now to explicate the argument of it in line with the account given so far. Kant introduces the synthesis of apprehension here as that whereby perception is possible and means by this "empirical consciousness of the intuition" (B160) but in stating this he effectively duplicates the view that ran throughout the A-Deduction account, that the synthesis in question has both an empirical and a transcendental form. The empirical form of it requires the use of the categories to unify the manifold but it is dependent upon a transcendental act that is itself the one that Kant goes on to describe. In this transcendental act the intuitions *of* space and time are described just as the passage just discussed above from A102 suggested they had to be. In the passage from B160 – 1n Kant speaks of a distinction between the forms of intuition and formal intuition. This distinction has been interpreted a number of ways.[12] However the way I am suggesting it should be interpreted is as a contrast between the manifold as unified by *a priori* intuitions (forms of intuition) and the conditions of possibility of the unity of the manifold (formal intuition). The latter, however, requires us seeing space and time as they were presented in §24 (and at A102), namely, as produced by the transcendental synthesis of imagination, which is why Kant closes the note by referring back to §24. The understanding determines sensibility through the transcendental synthesis of imagination and thereby produces the unity of the intuitions of space and time, a unity

---

[12] The most prominently influential reading of this distinction must be that by Henry Allison who essentially presents formal intuition as a categorial unity by contrast to the merely intuitive givenness of forms of intuition. For two versions of his account compare Allison, Henry E. (1983): *Kant's Transcendental Imagination: An Interpretation and Defense.* Yale University Press: New Haven and London, pp. 96 – 7 and the more recent formulation in Allison, Henry E. (2004): *Kant's Transcendental Idealism: An Interpretation and Defense.* Revised and Enlarged Edition. Yale University Press: New Haven and London, p. 115 and see my previous critical response to this view in Banham (2006) pp. 12 – 13.

that belongs to space and time themselves and yet which is possible for us to grasp by means of the concepts of reflection (and not the categories which is why Kant denies a role here for the "concept of the understanding").

Functions of Unity in Judgment and Intuition

Two questions remain to be discussed: firstly, how the account given responds to the contrasting views of Kantian intuition reported above and, secondly, how it enables a reconciliation of the passage from the B-Deduction at B160–61n with the assertion of the symmetry thesis in the Metaphysical Deduction. These two questions, whilst distinct, are related in the sense that the unity of intuition is involved in the understanding of them both. The description of the controversy over the determination of Kantian intuition given above showed that there is a question about the relationship between two apparently distinct criteria Kant provides for them. The criteria of singularity, which is introduced in the Transcendental Aesthetic when the intuitive nature of space and time are defended, is distinct from the general claim that immediacy attaches to intuitions as intuitions are directly connected to the basic experience of affection.

If we now trace the general view of intuition back to the account given above of the way that the manifold of intuition is genetically constructed for awareness then what is evident is that the immediate sense of intuition is a material one that is given to us as the basic experience that something is occurring at all. Within the isolation of the material sensible conditions of such givenness the simple notion of affection is the general sense that there is an experience. Without this there would be what Kant declares impossible, namely, "empty" moments. So the criteria of immediacy relates fundamentally to the sense that *there is* a "matter" of experience as we can say simply by pointing to our sense of being-affected. However, what we noted above, when we related the transcendental synthesis of imagination to the transcendental unity of apperception, is that the sense of such affection requires that the particular experience of affection be itself part of the continuity of moments being given. This is asserted clearly in the Anticipations of Perception as the basis on which we can discern a "magnitude" even within the basic experience of the "matter" of intuition. For this matter to be discerned as provided with a quantity however there has to be a way to be able to compare it with other forms of sensation so that its degree can be measured. This comparative possibility requires not only recourse to the concepts of comparison but also the ability to operationalize such concepts intra-temporally. The ground for such operationalization is thus the transcendental synthesis of imagination, which constitutes the possibility of the transcendental unity of apperception and which latter thus allows for awareness of the distinction between moments. With that entire package comes *formal intuition* or the unification of the matter of the intuitive manifold. This unified manifold enables us to cognise intuition as singular. So whilst the basic sense of the immediacy of intuition is primary in one way, the singularity criterion for intuition is primary in another. Genetically speaking the sense of immediacy is one that is the ground of all else as without it the formation of experience would have no matter on which to be built. However, cognition of the intuitive manifold only occurs when we have the unification of it into singular wholes and so in terms of cognition the singularity criteria (which brings with it what Caygill termed "formal principles") is prior.

The sense of the passage from B160–61n has been given above as showing that the view that space and time is itself internal to the intuition of them and not grounded on reference to the categories. But such an internal sense of the unity of space and time is a product of the action of understanding on sensibility that we can term the transcendental synthesis of imagination and which B160–61n rightly refers back to in §24. Now, in turning from the passage at B160–61n to that in the Metaphysical Deduction, it is helpful first to remind us what Kant takes a "function" to be since it is in terms of "function" that the symmetry thesis is stated there.

At A68/B93 Kant presents a succinct description of "function": "By 'function' I mean the unity of the act of bringing various representations under one common representation". So the term "function" describes how the multiplicity of representations is brought to a unity. If we now turn again to the passage at A79/B105 it is in order to apply this insight to the symmetry that Kant asserts here between the unity of intuition and that of judgment. The statement here is to the effect that giving of unity to thoughts in a judgment is provided by the same means that the unity of synthesis supplies as bringing together intuitions. What §19 asserts is that with objective judgments this occurs by means of the transcendental unity of apperception and what §24 shows is that the unification of intuition is likewise established by means of the transcendental unity of apperception, albeit in the latter case as a product of the synthesis of imagination. So when Kant goes on to claim that the unity is "the pure concept of the understanding" it is to this unity of apperception that he is primarily referring. Similarly the claim that the logical form of judgment brings about analytic unity is later described at B133 as dependent upon the synthetic unity of apperception. Finally the introduction of transcendental content into representations is described again as a product of synthetic unity which latter we can now see is the result of the transcendental synthesis of imagination.

So the symmetry asserted in the passage from the Metaphysical Deduction is not at variance with the claim made at B160 – 61n as effectively these two passages are different ways of asserting the same general point. This same point was made in the A-Deduction in a passage worth citing in conclusion:

> We entitle the synthesis of the manifold in imagination transcendental, if without distinction of intuitions it is directed exclusively to the *a priori* combination of the manifold; and the unity of this synthesis is called transcendental, if it is represented as *a priori* necessary in relation to the original unity of apperception. Since this unity of apperception underlies the possibility of all knowledge, the transcendental unity of the synthesis of imagination is the pure form of all possible knowledge; and by means of it all objects of possible experience must be represented *a priori* (A118).

Sidney Axinn
Symbols, Mental Images, and the Imagination in Kant

There is a well-known issue between those who take mental images to be pictures in the mind, and those who take symbols to do the job, to be the reference for certain concepts and for memory. Before behaviorism became established, the picture theory of the mind was widely assumed; since the decline of behaviorism the picture theory has had something of a return to serious consideration.[1]

Where would Kant be in this controversy? And, what is the status of mental images in Kant's use of imagination? He does say that we have "understanding to provide concepts, and sensible intuition to provide objects." [402]. The "objects", that sensible intuition provides, may easily be understood to be images. However, something else, the concept of a symbol had also been presented.

In a section in the *Critique of Judgment*, #59 "On Beauty as the Symbol of Morality," Kant gives us some analysis of symbols as well as his argument for the title of that section. An extended quotation will help make Kant's point about the important difference between a schematism and symbolism.[2]

> The effort to make a concept sensible
> is either schematic or symbolic. [351].
> In the schematic...an intuition corresponding to it is given. ...in the symbolic there is a concept which only reason can think. [if we provide a concept with objective reality by means of the intuition that corresponds to it] ...this act is called schematism. But if the concept can be exhibited only indirectly or mediately...this act may be called the symbolization of the concept [Note 31, 351].

After this we find,

> (the intuitive) can be divided into *schematic* and *symbolic* presentation: ...both are designations of concepts by accompanying signs. Such signs contain nothing whatever that belongs to the intuition of the object; their point is the subjective one of serving as a means of reproducing concepts in accordance with the imagination's law of association.

---

1 See the excellent article, "Mental Images," in the *Stamford Encyclopedia of Philosophy*, by Nigel J.T. Thomas. As usual, numbers in square brackets indicate the standard Berlin edition of Kant's works. References usually will be to the Werner S. Pluhar translation of the *Critique of Judgment*, (Indianapolis: Hackett: Publishing Co. 1987, with an occasional translation from J.H. Bernard's work, the *Critique of Judgment*, (New York: Hafner Publishing, 1951).

2 Square brackets are references to the Berlin Edition, as usual.

They are either words, or visible (algebraic or even mimetic) signs, and they merely express concepts[352]

In sum, "all intuitions supplied for a priori concepts are either schemata or symbols. Schemata contain direct, symbols indirect exhibitions of the concept." [352]. Shortly after this, Kant adds,

Our language is replete with such indirect exhibitions according to an analogy, where the expression does not contain the actual schema for the concept but contains merely a symbol for our reflection.[352]

As an example, after several qualifications, he says, "all our cognition of God is merely symbolic." (And to avoid the error of being literal about "cognition," the translator adds, "cognition by analogy.")

The relation between symbol and imagination needs attention. In the first Critique Kant had explained,

now it is imagination that connects the manifold of sensible intuition; and imagination is dependent for the unity of its intellectual synthesis upon the understanding, and for the manifold of its apprehension upon sensibility.[B164]

In the case of the natural numbers, there is no sense intuition, so we have no image of numbers. But, we deal with numbers quite well in terms of symbols. The significance of this (and other matters) has led Ernst Cassirer to insist that in addition to man being a rational animal, he is a symbolizing animal.[3] As Cassirer put it, "instead of defining man as an *animal rationale* we should define him as an animal symbolicum."[4]

In order to have a symbol make its proper reference, the imagination must assist. So the symbolizing animal needs imagination in an essential way.

Limits on Imagination

Consider this question: if we have never seen a certain object, nor have we seen the parts of it, can we imagine (have a mental image of) such an object? (The

---

3 Cassirer, Ernst (1944): *An Essay on Man*. New Haven: Yale University Press, and 1996:*The Philosophy of Symbolic Forms*, Four volumes. New Haven:Yale University Press. (Vol. Four trans. by John Michael Krois, edited by Krois and Donald Phillip Verene. There is also a three volume edition of *The Philosophy of Symbolic Forms*.

4 Cassirer's *Essay on Man*, p. 26.

history of philosophy has seen many authors consider this question.) Kant has given us his great *material* principle, "without material nothing whatsoever can be thought." (A232). The term, material, refers to whatever can give us an intuition. A symbol that may itself be observed can serve quite well. I call it a principle because Kant holds that it covers everything that can be thought.

This principle, that "without material nothing whatsoever can be thought," is so important to Kant that we must take it seriously and literally. To think of something there must be some material entity, something observable, with which to associate it or identify it. As mentioned above in the case of mathematics, if there is no other observation possible, an observable symbol will do to let us think. We can write *and then observe* the letters of the alphabet, mathematical symbols such as the symbol for square-root, for multiplication, division, for a range of numbers, an integral, etc. Therefore in mathematics we have the material (observable entities) with which to think. This is properly to be called a principle since it applies to anything "thought."

Considering Kant's material principle, what is the answer to the question, If we have never seen a certain object, nor its parts, can we have a mental image of such an object? The answer must be 'no;' without an intuition of material we cannot think such an entity. There is the additional matter: without any sense intuition, what word or words would we use to give ourselves, or anyone else, a description of such an object? We have seen horses and have seen wings, so we can (creatively?) put them together and imagine a flying horse (as Pegasus).

In the first Critique Kant had warned against empty objects, saying:

> we cannot...creatively imagine any object in terms of any new quality that does not allow of being given in experience. ..otherwise we should be resting reason on empty figments of the brain, and not on concepts of things. ...in a word our reason can employ as conditions of the possibility of things only the conditions of possible experience. [A770, B798-A771, B799]

Here Kant prohibits us from making epistemic claims that are not formulated in terms of experience. In this passage, as in other places, Kant insists that even a hypothesis must be based on possible experience to avoid being empty.

Mathematical Knowledge

The necessity of possible experience is repeated frequently by Kant, as in "all our knowledge falls within the bounds of possible experience." [A146]. If so, how is mathematical knowledge possible, since there is no experience of numbers and other entities of mathematics, such as the general idea of triangle, of the square root of a number, of a perfect circle, etc.?

The obvious answer to this question, as indicated above, is to deal with the symbols of mathematical entities, since there is no such entity as the numbers themselves to be experienced. The symbols are objects of experience, and so these objects make it possible to consider mathematics as an experimental science.[5] We observe the features of a step-by-step proof. When there is an error, we can point it out. When someone says that he can prove an assertion in mathematics, we say "show me," and the proof either can or cannot be presented for observation and verification. A mathematical proof has all the elements of an experiment in the other experimental sciences. *In mathematics the entities that are observed are symbols.* When we are given a definition of mathematics as the study of laws that are invariant under translation, these are laws stated as relations of symbols. The familiar laws in elementary physics are also so stated, for example, E=IR, and F=MA. All the laws in physics and chemistry are presented as relations of symbols. It is no insult to mathematics to note that the objects of observation there are symbols and relations between symbols. It was an important stage in the history of mathematics when it was "acknowledged that mathematics is not a theory of things but a theory of symbols."[6]

The argument, above, that mathematics is an experimental science in which the observables are symbols flies in the face of what good Kantian students know about Kant's view of mathematics. It is widely known that for Kant mathematical judgments are synthetic a priori. How then can they be experimental? In the First Critique Kant held that "All mathematical judgments, without exception, are synthetic" [B14], and followed this with "mathematical propositions...are always judgments a priori, not empirical."[B15]. But he then adds, "I am willing to limit my statements to *pure* mathematics, the very concept of which implies that it does not contain empirical but only pure a priori knowledge." [B15]. That leaves a lot of mathematics to be experimentally determined, as indicated above.

The usual index to the Critique of Pure Reason does not list 'symbol'. Symbols are not discussed in that work, but if Kant had worked out and used the material on symbols that we find in the third Critique, the section on schematism might have been much clearer. He does say, in the *Critique of Judgment,*

---

5 This view is taken from Axinn, S. (1973): "Mathematics as an Experimental Science". In: *Mathematics and Philosophy,* ed. Robert J. Baum. San Francisco: Freeman, Copper and Co.. Also, earlier in *Philosophia Mathematica,* Vol. V, No.1, 1968.

6 Cassirer, Ernst (1944): *An Essay on Man: An Introduction to a Philosophy of Human Culture.* New Haven: Yale University Press. p.60.

"our language is replete with such indirect exhibitions according to an analogy, where the expression does not contain the actual schema for the concept but contains merely a symbol for our reflection."[352]

If our language is "replete" with cases in which we have "merely a symbol," one would expect many more examples than just those in the beauty and morality case that he gives us.

One commentator on the history of mathematics wrote that "Immanuel Kant, for instance, shows no familiarity at all with the advanced mathematics of his times."[7] Of course, where Kant's interest was in questions such as "How is pure mathematics possible," and "How are a priori synthetic judgments possible," these call for a "transcendental" inquiry. That is quite apart from a pursuit of the detailed questions in the science of mathematics. He has insisted that "...pure mathematics is distinguished from applied mathematics..."[8] Pure mathematics is in the area of metaphysics, applied math is an empirical subject, and depends on observation of symbols.

In a helpful footnote, Kant notes that when something is symbolic, it is "a presentation in accordance with a mere *analogy*." [352, note 32].

How is Beauty a Symbol of Morality?

Kant starts his answer to the question of the relation of beauty to morality by asserting that we regularly do "refer the beautiful to the morally good" [353]. That sentence adds that we require others to also do so, and that we feel "ennobled" by doing so. It seems to me that there is a stronger argument for Kant's view of beauty and morality than he gives, and I'll present it.[9]

The relationship between beauty and morality loses its strangeness if we start with a remark that Hannah Arendt made in one of her last works, published as lectures on Kant's political philosophy.[10]

---

7 Bochner, Salomon In: *Dictionary of the History of Ideas*, Vol. III, editor in chief, Philip Weiner, p. 179. The eminent mathematicians of that time, as Bochner mentions, were, Leibniz, Jacob and John Bernoulli, Clairaut, Eulelrm, d'Alambert, de'Maupertiuis, and Lagrange.

8 *Groundwork for the Metaphysics of Morals*, (2002) [4:411], in the translation by Thomas E. Hill, Jr. and Arnulf Zweig. Oxford: Oxford University Press, p.212, footnote.

9 The view to follow is taken from Axinn, S. (1991): "On Beauty as the Symbol of Morality". In: Fuinke, G. (Hrsg.): *Akten des Siebenten Internationalen Kant-Kongresses*, Kurfürstliches Schloss zu Mainz, 1990. Bonn: Bouvier.

10 Arendt, Hannah (1982): *Lectures on Kant's Political Philosophy*, edited by Ronald Beiner. Chicago: The University of Chicago Press.

"The only objects that seem purposeless are aesthetic objects, on the one hand, and men, on the other; you cannot ask quem ad finem?- For what purpose?- Since they are good for nothing."[11]

Of course Arendt knew the Kant texts, and knew quite well the idea of a purposeless purpose. Kant takes an object to be beautiful if it has what he calls "subjective purposiveness in the presentation of an object, without any purpose...and hence the mere form of purposiveness..." [221]. His view of beauty and of the notion of "the form of purposiveness" involves complications beyond present needs.

A few quotations may suffice. Kant finds contradictions and inconsistencies in the matter of beauty, and he trumpets them rather than ignoring or pretending to make them consistent. "...judgment finds itself referred to something that is both in the subject himself and outside him..." [353] "...the theoretical and the practical power are in an unknown manner combined and joined into a unity."[353]. "We present the subjective principle for judging the beautiful as *universal*, i.e., as valid for everyone, but as unknowable through any universal concept."

For this presentation we simply follow Arendt in noting that art seems to have no purpose, and in the case of humans, Kant has noted that "we cannot see why people should have to exist." [378].

Arendt has insisted on a significant point: the universe could get on without aesthetic objects, and it could also exist without humans. It could exist without humans collectively, and also exist without any particular human, even the most gifted genius. Without Plato, or Euclid, or Einstein, without Beethoven or Christ or Confucius...we know that nature would continue without any of them, or of us. Neither art objects nor humans are necessary for the universe: Arendt insists that we are "good for nothing."

The distinctive feature of art objects is beauty, and it seems to have no "objective" purpose. Such a purpose would be one that exists independently of humans. The distinctive feature of humans is morality, again with no "objective" or external purpose. What follows from this view of Arendt's? Suppose we add that humans *know* they exist for no purpose of nature.

Humans realize that they exist for no purpose, and yet still want to see themselves as significant. Can such a contradiction exist? Is there any model of such a contradictory state of affairs? A desperate question; can there be something that has no purpose and yet does have some sort of purpose? Fortunately, beautiful

---

11 Arendt, Hannah (1982): *Lectures on Kant's Political Philosophy*, edited by Ronald Beiner. Chicago: The University of Chicago Press.p.76.

objects fit this requirement exactly. Following Kant's definition of beauty [236],[12] we have a ready explanation of the human fascination with the beautiful. Beautiful objects provide empirical examples of exactly what is so intensely desired, objects that seem to have no purpose and yet are extremely important. If there are such objects in the world, and if they are universally recognized as such, then we have a defense of our own possible significance...by analogy. This analogy is why beauty can symbolize morality. Neither is needed for nature to exist, yet we take them each to be supremely important. One should note that the above interpretation makes the relation between beauty and morality an analogy, not a symbolic matter. Of course, this interpretation drawing on Arendt is not Kant's.

Conclusion

Kant has distinguished signs, schemata, symbols, and imagination. Presumably he would be quite satisfied with the work of Ernst Cassirer in his four volume, *The Philosophy of Symbolic Forms*, and his *Essay on Man*. Kant has told us that "language is replete" with the use of symbols. [352]. Cassirer's work gives the argument and some details of how man uses symbols. Cassirer applies his work on symbols to art, language, myth, and other aspects of culture, and to recent work in mathematics (in his time).

Signs are related to particulars, symbols can be general, as in the difference between proper nouns and common nouns. While Kant uses the notion of imagination frequently, he is not very clear on just what it is, and other authors in this collection should be consulted on the matter.[13] To return to the earlier question of the status of the picture theory of mental image, Kant is also not as clear as one might wish. While he points out that imagination requires sensations, the exact result of sensations in the mind is not developed. He does say, as noted above, that intuitions provide sensations that give us objects. What sort of ob-

---

12 *As Kant puts it in the Third Critique, "Beauty is an object's form of purposiveness insofar as it is perceived in the object without the presentation of a purpose."* [236] What is meant by "form" without "presentation"? Presumably a sense of purpose but with no particular specification of it. Here, using words as symbols avoids having a bare symbol and a repetition of it contradicting the first.

13 In Kant's defense it may be noted that imagination is widely considered difficult to define. For example, a prominent contemporary, Kendall Walton, author of *Mimesis as Make Believe* (Cambridge: Harvard University Press) 1990, has said that he can't define imagination because there are too many dimensions.

jects is not analyzed in detail. He does use the term "image," frequently, e. g., five times in [234], Pluhar's translation.

Some aspects of the subject had already been considered in the First Critique. "Indeed it is schemata, not images of objects, which underlie our pure sensible concepts."[A141] And then he added, "No image could even be adequate to the concept of a triangle in general."[A141]. "...the image is a product of the empirical faculty of reproductive imagination: the schema of sensible concepts, such as figures in space, is a product...of pure a priori imagination." [142].

Kant's view of the role and frequent use of symbols, however, is a contribution that should be more widely credited.[14] Using Kant's view of symbols to reconsider material in the first two Critiques is a task still ahead.

---

14 Even Cassirer, a Kant scholar himself, does not mention Kant's work on symbols, as far as I know. In his *The Philosophy of Symbolic Forms* there are many references to Kant. And he gives Kant's "schematism" major attention, but I have not found any references to Kant's use of "symbol."

Bernard Freydberg

Functions of Imagination in Kant's Moral Philosophy

It is not possible to delineate the functions of imagination in Kant's moral philosophy without traversing the way synthesis pervades the *Critique of Pure Reason*. This paper's orientation is "continental", in the most general sense that it lets these texts be its guide rather than assumes that our "wiser" contemporary philosophical advances allows us to use Kant for other commitments and views. As there is little scholarship that escapes both the snares of Anglo-American Kant scholarship, which downplays imagination, and the Heideggerian strain of continental scholarship, which locates imagination within the quest for a fundamental ontology, this paper is content to follow the Kantian text where it leads and in its own terms. Although such treatment is rare in these times, I am confident that many will find it useful and clarifying.

The passage that serves as the anchor occurs at A78, B103:

> Synthesis in general, as we shall hereafter see, is the mere result of the power of imagination [*Einbildungskraft*], a blind but indispensible function of the soul, without which we should have no knowledge whatsoever, but of which we are scarcely ever conscious.

Understanding is distinguished from imagination in that the former brings the synthesis *"to concepts"* (A78, B103). The threefold of sensibility (immediate relation through intuition), imagination, and understanding, of course does not imply a temporal sequence. Their togetherness has always already occurred.

The paper will be divided into four sections. (I) An account of the Transcendental Aesthetic and Analytic of the *Critique of Pure Reason*; (II) an account of the Transcendental Dialectic of the *Critique of Pure Reason*; (III) an account of the thoroughly synthetic character of the *Critique of Practical Reason* (the longest section); (IV) the enactment in actual experience of moral principles, in terms of the *Metaphysics of Morals*, in which I address a few of Kant's many anachronistic views. My conclusion will gather up what preceded, and add a few brief words.

1 Account of the Transcendental Aesthetic and Analytic of the *Critique of Pure Reason*

I do not deny that exegetical challenges remain when one interprets this passage in its strict sense, as I do. That is, I interpret *"die blosse Wirkung der Einbildung-*

skraft" (the mere result of the power of imagination) as claiming that *all* synthesis is the result of imagination; the *bloss*, in this context, indicates that *only* imagination generates synthesis. Of course Kant scholarship would neither be so extensive nor so hotly contested if exegetical matters could be so straightforwardly settled. In §15 and§16 of the B-edition Transcendental Deduction, Kant speaks of combination as implying synthesis, and ascribes both to understanding. In the *Critique of Practical Reason*, Kant appears to take special pains in order to exclude imagination from moral considerations. In the typic, which concerns the application of the moral law to instances in the realm of appearances, he ascribes this function to "understanding (and not imagination)".

However, if we allow these passages to determine the meaning of the Kantian text definitively, we have difficulties that are insurmountable rather than needing some minor reconciliation. Regarding the B Transcendental Deduction, I depart not only from those Anglo-American scholars who find vindication for their general Kant-interpretations in the absence of the "psychologistic" imagination that dominated the Deduction in the A-edition. I also depart from Heidegger, who also found a dramatically diminished role for imagination in the B-edition for reasons described above. The task of the Deduction is to justify the application of the Pure Concepts of the Understanding (that is the categories) to the objects of sensation that have an origin heterogeneous to the categories.

One can count words and so find fewer instances of "imagination" in the B deduction than in its "A" counterpart. However, a superior and more decisive text settles the question unmistakably in my view. §24 of the B Deduction is titled *"The Application of the Categories to the Objects of the Senses in General."* That is to say, in this section the Transcendental Deduction both properly takes place and finds its completion; every earlier section is in service to this one. As I have interpreted this matter elsewhere[1], the absence of imagination in §16-§20 signals two points: (1) if the only two faculties under discussion are understanding and sensibility, and understanding is the spontaneous faculty (sensibility is by nature receptive, passive), then understanding is the sole faculty to which synthesis can be ascribed, and (2) the two-faculty presentation of the objective deduction is very likely a response to the Garve-Feder review that likened Kant's first *Critique* to Berkeley's principles, based upon what they saw as an outsized role of the subjective side, that is, of imagination.

---

[1] Freydberg, Bernard (1995): "Concerning 'Syntheses of Understanding' in the B Deduction".In: *Proceedings of the Eighth International Kant Congress*, Volume II, Part 1. Memphis: Marquette University Press 1995, 287–294

A word on "faculty" is in order, a notion that has provoked much attention, often negative, on Kant's thought. The German is the gerund *Vermögen*; the verb *vermögen* means to be able, to have power. Its grammatical rendering as a substantive does not transform it into a real entity, any more that rendering the verb *denken* (to think), as *Denken*, thinking, posits a real entity apart from the activity to which it points. To speak then, as so many have, of a "faculty psychology" in Kant is doubly misleading. There are no substantive "faculties," nor is there a "psychology" in anything like our sense of the word. When we use it today, we mean either the science of behavior generally, or some branch of psychoanalysis. When Kant employs the word, he does so in the sense of a doctrine of the soul (the "I think"). As the "I think" has nothing manifold in it, strictly speaking there is no such doctrine; there is only the unification of the whole of what is given in inner sense as guided by the idea of the unity of all experience. To speak of a faculty, imagination, providing synthesis is to speak of an active ability to run through a manifold and to hold it together.

At the beginning of §24, Kant seems to complicate matters by writing of two distinct syntheses, an intellectual synthesis that occurs on the level of apperception alone "so far as such [*a priori*] knowledge rests upon the understanding." (B150) However, this intellectual synthesis—I say it occurs nowhere else but in speech, as an abstraction from the one entire synthesis of imagination—is emphatically not the concern of §24. Against my view, one could argue that Kant calls both syntheses "not merely as taking place *a priori*, but also as conditioning the possibility of other *a prio*ri knowledge. (B151) But immediately thereafter, he makes clear that the figurative synthesis *alone* occupies this section, and is called in order to distinguish it from the merely intellectual combination, the *transcendental synthesis of imagination*." (B151, emphasis in original).The figurative synthesis, which includes "the manifold of sensible intuition, which is possible and necessary a priori" is its aim. Kant gives a decisive characterization to the figurative synthesis as directed to the original synthetic unity of apperception: with emphasis, it is called *"the transcendental synthesis of imagination."* This synthesis gathers all that has preceded, as it prepares the way for what will follow.

The exegetical difficulties presented by this problematic distinction present little difficulty for the *Critique of Pure Reason* in my view. However, matters will deepen when confronting the *Critique of Practical Reason*. For the first *Critique*, one needs only to bear in mind that the synthetic a priori judgments that constitute the possibility of both experience and the objects of experience are the four principles (*Grundsätze*) of the Pure Understanding (A161, B200, B294). As the Doctrine of Elements progresses, it becomes clear that each of the separately treated elements cannot exist apart from their togetherness in the principles.

The Transcendental Aesthetic takes place by virtue of a most problematic double abstraction: (1) it abstracts entirely from understanding with its concepts, and (2) it abstracts from everything empirical, and concerns itself only with pure intuition. However, one need not be a very shrewd exegete to notice the intrusion of supposedly excluded material, that is, magnitudes, "empirical reality," etc. Still more problematically, there resides an almost unbearable tension at the heart of the notion of "pure intuition.' "Pure" is said to mean "non-empirical." But intuition—human intuition, that is, *our* intuition—is said to be restricted to sensibility, that is, to the empirical realm. So how can pure intuition even be thought? (In my 1973 seminar, Lewis White Beck opted for "pure form of intuition" and we were instructed to make that replacement. By contrast, in his 1977 course John Sallis interpreted it to mean "the pure element in every empirical intuition." I found greater advantage in the latter interpretation, for reasons that will become clear).

The section titled "Transcendental Analytic" seems to have thought, the other side of human knowing, as its concern. "Thoughts without content are empty, intuitions without concepts are blind," Kant's famous saying reads (A51, B75). But perhaps the emphasis that this phrase has received is misplaced; at the very least, I say that it has functioned often in a most misleading manner. For one thing, the double abstraction on the side of logic yields what Kant calls "general logic," a term that corresponds to our "formal logic." General logic has only one function for theoretical knowledge: to exclude any knowledge claims that are internally self-contradictory. These are few, and these few are easily detected. Once one transgresses the limits of theoretical knowledge, matters become much more interesting, especially for practical philosophy. But this is not the concern in the Analytic.

Rather, the so-called abstraction is—to speak awkwardly—compromised in the Analytic. The logic presented in it, Transcendental Logic, emphatically does not exclude sensibility, but, rather, includes it as an essential feature. For purposes of directness, I avoid the interpretive thicket surrounding the Metaphysical Deduction of the general-logical categories in order to move to the Pure Concepts of the Understanding, the categories proper. These can be presented in the following formula: the general-logic categories + *pure intuition* = the Pure Concepts of the Understanding. Since time is the sensible condition of all intuition, that is, both outer and inner, time belongs intrinsically to the Pure Concepts of the Understanding.

Why did I write of a "so-called" abstraction? Once again, the order of presentation does not reflect the actual order of the judgments in question. Every judgment made by every human being—including analytic judgments, as is later shown—presuppose the Pure Concepts of the Understanding.[2] As the true Kantian position begins to take shape, so too does the nature of these categories proper: they are the product of an original synthesis, anterior to their elements in the sense that their elements are nothing whatsoever apart from this prior synthesis.

To broach a much-discussed subject all too quickly, this is why Kant writes of three syntheses in the A-edition Transcendental Deduction, as if the three could be separated in deed as well as in speech—which they cannot. The Synthesis of Apprehension in Intuition presupposes the Synthesis of Reproduction in Imagination, which in turn presupposed the synthesis of Recognition in the Concept. This skeleton of the A Deduction occurs in a section titled "The A Priori Grounds of the Possibility of Experience," and so does not constitute the Deduction proper. The "Grounds" consist of the three syntheses thought in their unity under transcendental apperception. After declaring that "only the productive synthesis of imagination can take place *a priori*," (A118) Kant writes crucially:

> The unity of apperception in relation to the synthesis of imagination is the understanding; and this same unity, with reference to the transcendental synthesis of the imagination, the pure understanding. In the understanding, there are then pure a priori modes of knowledge [Erkenntnisse] that contain the necessary unity of the pure synthesis of imagination in respect to all possible appearances. These are the categories, that is, the pure concepts of understanding. (A119)

This is the very same transcendental and productive synthesis of imagination that occurs in §24 and that was discussed above. There have been some recent efforts—unjustified and, in my view, unfortunately influential—to read pure imagination out of Kant.[3]Consider the Transcendental Deduction of the categories without pure imagination: the connection to intuition could not be established. Think about the schemata, pure procedures of imagination one and all, without imagination. The schemata provide "sense and significance" (*Sinn und Bedeutung*) to the categories; without them, *a fortiori*, the categories would be with-

---

2 See especially B 133n: "...only by means of a presupposed synthetic unity can I represent to myself the analytic unity." This holds for all analytic judgments, which one and all occur under the transcendental synthetic unity of consciousness.

3 For example in Paul Guyer's introductory general essay in *The Cambridge Companion to Kant* (Cambridge: Cambridge University Press, 1992), he does not so much as mention imagination. Let us consider the consequences for Kant's thought if imagination were excised: (1) synthesis would be excised, and with this the synthetic a priori judgments that are so vital; (2) the Transcendental Deduction of the Categories (either the A or B edition); (3) the schematism, by virtue of which alone the categories receive sense and significance (*Sinn und Bedeutung*); and therefore (4) the Principles, which make both experience and objects of experience possible.

out sense and significance. Thus, reading imagination out of the *Critique of Pure Reason* would be akin to reading Beatrice out of the *Divine Comedy*, or (perhaps in another way, but an analogous one) to excising Lucifer from *Paradise Lost*. From top to bottom, the *Critique of Pure Reason* presents synthesis at every turn—indeed, without synthesis, nothing occurs at all.

2 From Transcendental Dialectic to the *Critique of Practical Reason*

The Ideas of Pure Reason, generated by the Categories of Relation that have abandoned their connection to pure intuition, cannot possibly provide a ground for knowledge of experience appropriate to a being like us, that is, a being that is bound to empirical intuition. The Idea of the Soul arises from the illicit removal of the sensible condition from the Category of Substance; likewise, the Idea of the World arises from the detaching of the Category of Cause from its sensible condition; the Idea of God arises from the Category of Community's release. They are one and all, concepts of the unconditioned (A322, B379). The judgments in which they occur are one and all both synthetic and *a priori*. In their own peculiar way, they serve the theoretical interest by providing subjective unity to our empirical knowledge. The instance among them that concerns us here occurs in the Third Antinomy, where the principle of non-contradiction enters the discourse. The relation of synthesis to this apparently analytic incursion requires the closest attention.

Both the thesis, which claims the necessity of assuming another causality besides natural necessity, namely freedom, and the antithesis, which claims that there is no freedom but only natural necessity, are synthetic and *a priori*. Both include the idea of the world, and both define the world as the absolute totality of conditions. Both grant the unifying role of this idea. Why does this principle occur in this context, and how exactly does it function? The most precise answer to the first part of the question is found in a footnote in the preface to the B-edition:

> To *know* an object I must be able to prove its possibility, either from its actuality as attested by experience, or *a priori* by means of reason. But I can *think* whatever I please, provided only that I do not contradict myself, that is, provided that my concept is a possible thought. (Bxxvi)

Thus, all of the judgments in the Dialectic can be thought. That, indeed, is the difficulty they present: since they are extensions of the categories through which we must think any object at all, the illusion of having the capacity to provide knowledge of the supersensible region is unavoidable. However, when the issue concerns not knowledge but a kind of *access* to the supersensible, matters become more complex. This access opens precisely on account of the negative role of general (formal) logic and its principle of non-contradiction. That is, the three Ideas and the synthetic *a priori* judgments in which they occur are one and all *thinkable*. One can think the immortality of the soul, spontaneous (free) causality in the world, and the existence of God without fear of being contradicted. Of course, one can also think the mortality of the soul, the bond to absolute necessity of the world, and the non-existence of God without fear of being contradicted.

These judgments, as one and all synthetic, are one and all the product of imagination. They differ from their counterparts in the analytic in that they are unschematized, that is, they lack a connection to pure intuition (or any intuition whatsoever). However, it is not surprising that, as both synthetic and *a priori*, there occurs an intrinsic affinity to their theoretical counterparts. In a somewhat clumsy, but nevertheless appropriate characterization, Kant calls an intermediary between an idea and the manifold of understanding that it serves to unify a *schema-analogon*. The constitutive unity that the principles provide in order to make experience possible as established in the analytic is paralleled in the dialectic by the regulative unity provided by the judgments incorporating the ideas. In both cases, an intermediary that partakes of both heterogeneous stems is required, and in both cases it is the function of imagination to answer that need.

Once again, however, the order of presentation does not reflect the actual order in which these judgments always already function. Appearances become objects of experience through the principles, and these objects of experience become unified by the ideas—always already. For example, my breakfast of corn flakes and my attendance the next evening at a jazz concert are both mine as a result of the unity provided to apperception on the level of theory, but require further unification in order to be part of a *whole*. The latter unity has always already been affected by the schemata-analoga of imagination. However, this latter unity occurs in a register unlike that of the theoretical one.

This is hardly the occasion to revisit the Third Antinomy and the thicket of issues contained therein. Instead, I shall focus upon the schema-analogon of freedom. Adhering strictly to the logical side of the Antinomy, Kant's conclusion could hardly be more modest: freedom as spontaneous causality is admitted as possible only for the slimmest of reasons, namely because its reality does not entail a contradiction. Nevertheless, one can say the following with some anticipatory confidence: just as causal necessity applies, by means of the idea of the world together with the schema-analogon, to the whole of external experience as it extends through all time, the idea of freedom, together with that same idea of the world and its schema-analogon, applies to the whole of inner experience. The *Critique of Practical Reason* justifies this confidence.

3 The Thoroughgoing Synthetic Character of the *Critique of Practical Reason*

Kant comments upon the systematic nature of the whole of human knowledge, and its order of presentation. The passage is crucial, and I cite it in full:

> When it is a question of determining the origin, contents, and limits of a particular faculty of the human mind, the nature of human knowledge makes it impossible to do otherwise than begin with an exact and (as far as is allowed by the knowledge we have already gained) complete delineation of its parts. But still another thing must be attended to which is of a more philosophical and *architectonic* character. It is to grasp correctly the *idea of the whole*, and then to see all those parts in their reciprocal interrelation, in the light of their derivation from the concept of the whole, and as united in a pure rational faculty...Those who are loath to engage in the first of these inquiries...will not reach the second stage, the synoptic view [*Übersicht*], which is a synthetic return to that which was previously given only analytically (*Akad.* 10)

Beck's translation of *Übersicht* as "synoptic view" is clearly a reach beyond its literal sense. However, it is not only defensible, but praiseworthy, when considered in light of its counterpart in the *Critique of Pure Reason* at A97: "As sense contains a manifold in its intuition, I ascribe to it a synopsis (*Synopsis*). But to such a synopsis a synthesis must always correspond..." Kantian language is said to be difficult, even notoriously so, but at its basis it could hardly be clearer. Synthesis, from the Greek *sun-thesis:* to put together. Synopsis, from the Greek *sun-horaō:* to see together. Beck's "synoptic view": to see (or to survey) the whole together with its interrelated parts.

This theme has occupied my discourse all along: the order of presentation does not represent the actual order of occurrence. The parts are delineated serially, but they always already occur in tandem. This passage from the *Critique of Practical Reason* does more than merely confirm what has been said so far; it declares in unmistakable terms and without exception that the ultimate philosophical act is always the prior philosophical act, namely synthesis—putting-together. The "alreadiness" points to the pure *a priori* nature of such synthesis. The synthesis reaches its highest point in the moral law.

In my book on Kant's moral philosophy,[4] I employed a way of interpretation that has found a place in continental philosophy that has been characterized as reading a text *against itself*. An exemplar of such interpretation can be found in John Sallis' *Spacings—of Reason and Imagination in the Texts of Kant, Fichte, and Hegel*. Sallis points out an early passage in which Kant writes:

> Pure reason is, indeed, so perfect a unity that if its principle were insufficient for the solution of even a single one of all the questions to which it itself gives birth we should have no alternative but to reject the principle... (Axiii)

However, this unity runs aground in the dialectic where (as indicated above) there can be nothing within reason itself that can settle the disunion between the theses and the antitheses.[5]

In the second *Critique*, Kant takes special pains to exclude imagination from the discourse, which he insists is to be governed by reason alone. In the typic of Pure Practical Judgment, which has a role analogous to that of the schematism of the first *Critique*, Kant seems to build a firewall between them. After apparently equivocating (Kant writes of "the schema [if this word is suitable here of a law itself,"] he writes:

> A schema is a universal procedure of the imagination in presenting a priori to the senses a pure concept of the understanding which is determined by the law [i.e., the a priori laws of nature]...But to the law of freedom...and consequently to the concept of the absolutely good, no intuition and hence no schema can be supplied for the purpose of applying it *in concreto*. Thus the moral law has no other cognitive faculty to mediate its application to the objects of nature than the understanding (not the imagination); and the understanding can supply not a schema of sensibility but a law. (*Akad.* 69)

However, one must ask: what is the epistemological status of this law supplied by the understanding? Instead of a schema, Kant calls it "the *type* of the moral law" (*Akad.* 70). When we hear type, we would do well to hear its late fifteenth century Latin meaning as "form," though it also meant figure, image, or kind. There is a clear distinction between "schema" and "type." The schemata are what might be called especially sapient procedures: they contain within them-

---

4 *Imagination in Kant's Critique of Practical Reason.* (Bloomington and Indianapolis: Indiana University Press, 2008).

5 Sallis writes concerning the fissure within pure reason: "First, reason is essentially one; and yet, in misunderstanding it can be set against itself, can become twofold (as Kant will most graphically show in the antinomy of pure reason.)" See Sallis, John (1987): *Spacings of Reason and Imagination in the Texts of Kant, Fichte, and Hegel.* Chicago: University of Chicago Press, 11.

selves the unerring ability to exercise judgment "correctly." There can be no mis-application of the schema of quantity, although mismeasurement is clearly al-ways possible; while determinations of, for example, causality require an empir-ical manifold, the succession of time always occurs. Nature is nothing other than that which occurs on the basis of the principles, which one and all vouchsafe its possibility.

By contrast, a gap does exist between the type and the nature with which it connects. The schemata, in giving the categories sense and significance, are cen-tral to the "Analytic of Understanding" to which the "proud name of an ontolo-gy" must yield. "Being" means the objectivity provided by principles under the transcendental unity of apperception. By contrast, the moral law provides not "being" but expresses an "ought." The type is therefore "not as smart" as a sche-ma. It governs not principles, but maxims of nature—whether they are in accord or not in accord with the moral law's demand for universality:

> in cases where the causality of freedom is to be judged, natural law serves only as the type of a law of freedom, for if common sense did not have something to use in actual experi-ence as an example, it could make no use of pure practical reason in applying it to that experience. (*Akad.* 70)

For Kant the distinction between a schema and a type plays an important meth-odological role in Kant's practical philosophy. The a priority of the type prevents an empiricism of practical reason, but more to the point here:

> the same typic also guards against the mysticism of practical reason, which makes into a schema that which should serve only as a symbol, i.e., proposes to supply real yet non-sen-suous intuitions (of an invisible kingdom of God) for the application of the moral law, and thus plunges into the transcendent [*Überschwengliche hinausschweift*].[6] (*Akad.* 70–71)

However, this distinction opens out into a series of complications for moral phi-losophy that can be addressed only by means of imagination. The schematism of judgment accounts for the bringing together of the two heterogeneous branches of human knowledge that has always already taken place. The typic rules out *a priori* such bringing together.

The principles one and all incorporate their respective schemata. The two most prominent and promising for providing examples are those of substance and of cause and effect. Their schemata, respectively, are "the permanence of the real in time" (A144, B183), and "the real upon which, whenever posited,

---

6 "Careens into exaltation" would be a literal translation.

something else always follows" (A144, B183). In both cases, "the real" refers to a "something in general" given to empirical intuition. However:

> the schema of a pure Concept of Understanding can never be brought into any image what-
> soever. It is simply the pure synthesis, determined by a rule of that unity, in accordance
> with concepts, to which the category gives expression It is a transcendental product of
> imagination, a product which concerns the determination of inner sense in general accord-
> ing to conditions of its form (time), in respect to all representations...(A142, B181)

Elsewhere, I have characterized the positive outcome of Kantian theoretical philosophy in the following manner: the principles provide a field upon which representations can appear and become objects or objective events for us. For example, the chair in my office is an instance of substance as phenomenon. One can walk around it from left to right or from right to left, peer over it, etc., and the order of such survey is a matter of indifference.[7] This is not the case with regard to my dropping a five-pound weight from four feet above on my desk and the loud sound made by contact. One never hears the sound occurring first, then the weight moving upward to my hand.

The schemata supply the means by which we "spell out," or organize our appearances so that they acquire objective validity. Of course, there are a great many cases where the causal connection is in doubt. Nevertheless, the schema of the principles of causality governs the research, for the most part implicitly. Research in medicine and nutrition often grope for causal connections, sometimes for years in vain. This is true of much cancer research. However, there are notable successes, for example, concerning the connection of smoking with cancer and heart disease, and the discovery of vaccines that virtually eliminate the scourge of polio. Most casual opinions concerning causality lack the connection demanded by the schema, e.g., prayer in schools causes greater abstinence in teen sexual activity or that political liberalism causes greater respect among men for women. The schema embedded in the principle provides the testing procedure according to which appearances are either judged to be objectively valid, objectively invalid, or indeterminate. The power of this procedure rests upon the connection of the schemata to pure intuition, that is, to time.

This connection is entirely absent from moral judgment. To say, as Kant does, that a type is a symbol and not a schema abrogates the connection to pure intuition. The symbolic relation of type to instance announces a gulf between them; their relation is indirect. One might argue that the theoretical rela-

---

7 Regarding the Mathematical Principles, it has a certain size, that is, extensive magnitude; regarding its color and the color's intensity, it has an intensive magnitude.

tion of schema to instance is also indirect, but a homogeneous element does obtain between the concept and the instance, namely pure intuition (time). On the theoretical level, where the status of objects is in question, the issue of motivation does not come up. Things do not have motives. Persons do. In persons, two general motives compete: self-love that aims at happiness and morality that aims at goodness. Both motives are always at play in persons.

This play enforces a clear limit to the evaluation of a person's moral judgment:

> It is in fact absolutely impossible by experience to discern with complete certainty a single case in which the maxim of an action, much as it may conform to duty, rested solely on moral grounds and on the conception of that duty. (*Akad.* 406–07)

Even in the most apparently pure of maxims of actions, the motive of self-love can never be entirely ruled out.

The interpretation of actions themselves in terms of their morality grows ever more complicated. Beginning from the categorical imperative that expresses the moral law, one proceeds to maxims that satisfy the demand for universalization. The intentions that conform to these maxims pass the test, those that do not fail the test. Both the pure moral law expressed in the categorical imperative and the maxims derived from it are synthetic a priori propositions. There is no denying that imagination, the function that enacts all synthesis, is at work in these judgments. Given Kant's claim that practical philosophy concerns itself with reason alone, the syntheses here *may* be the proper location of what Kant called intellectual synthesis in §24 of the first *Critique*. However, his more extensive account seems more consistent with the figurative synthesis:

> The figurative synthesis must...in order be distinguished from the intellectual combination, be called the *transcendental synthesis of imagination. Imagination*, is the faculty of representing in intuition an object that is *not itself present.* (B 151)

I say "more consistent" because *actions* rather than mere objects appear in intuition when considered in light of morals. Every appearance surely conforms to natural causality. In the same way, so too does every action. Nevertheless, causality through freedom allows for another interpretation, which Kant characterizes as *lawfulness*—an interpretation that one might call "over and above" that of appearances in terms of natural causality:

> From this point of view, a rational being can say of any unlawful action that he has done that he could have left it undone, even if as an appearance it was sufficiently determined in the past and thus far was inescapably necessary. For this action and everything in the past

which determined it belong to a single [*einzig*] phenomenon of his character, which he himself creates, and according to which he imputes to himself as a cause independent of all sensibility the causality of that appearance. (*Akad.* 98)

A lawful action—what does this entail? It entails nothing other than action performed in accordance with the moral law, itself a synthetic *a priori* proposition. Imagination has been shown to be the source of all synthesis and of representing what is not itself present. The maxims themselves may be implicit, and, as indicated above, may be complex, that is, may be chosen from duty but may nevertheless have a measure of self-love in them. Nevertheless, just as the criterion of universality implies the syntheticity of both the law and of the maxims conforming to them, the actions that result from them are marked by their synthetic origin—which is to say, by their ultimate relation to imagination.

4 Enactment in Actual Experience of Moral Principles in Terms of the *Metaphysics of Morals*

At this point, the gaps that are inherent to actual practical judgment become manifest, and the many odd stances Kant takes in particular cases become if not explicable, then at least plausible. In cases where the action is in accord with duty and thus possibly done from duty as well, we can at least strongly suppose that a genuinely moral intention is motivating that action. Such actions include, following Kant's examples in the *Foundations of the Metaphysics of Morals*, those that preserve one's life in the face of great hardship, those that are truth-bearing when convenience (happiness) might lead one to lie, those that are kind without benefit to the bestower, and those that develop talents for the good of humanity alone. There are a great many actions "in the middle," so to speak. Although Kant does not treat them as such, his text demands such an inference. All four of the above groups contain actions that are performed only in accord with duty and so are motivated by a mixture of morality and the desire for happiness. Those actions from maxims that are contrary to duty are so regarded because no maxim from which they follow can be universalized.

Kant's own application of these principles, however, has many counterintuitive if not ridiculous results.[8] I will use masturbation (a transgression he cannot

---

8 In *Kantian Ethics* (Cambridge: Cambridge University Press, 2008), Allen Wood devotes a

even bring himself to call by name) as my example, which he regards as more morally reprehensible than murder:

> That such an unusual use (and so misuse) of one's sexual power is a violation of duty *to oneself*, and indeed, one which is contrary to morality to the highest degree occurs to everyone immediately, along with the thought of it, and stirs up an aversion from this thought to such an extent that we consider it indecent to call this vice by its proper name. This does not happen in the case of self-murder, which we do not hesitate in the least to lay before the world in all its heinousness... (*Akad.* §7, 424)

His *Metaphysics of Morals*, from which this passage is drawn, features frequent interludes of casuistic questions in which moral dilemmas are raised. For example, Kant wonders whether the suicide of one who knows with certainty that madness contracted from a mad dog's bite would put others around him in danger has committed a moral wrong. But no such set of casuistic questions accompanies the matter addressed above. Can we merely attribute this lack of parallelism to the vast difference in mores between Kant's culture and ours? Or might we wonder whether, for example, a Catholic priest who finds himself sexually attracted to young boys might do better to indulge "self-pleasuring" rather than use his position of power to exploit the boys under his care?

Only toward the end of the Ethical Doctrine of Elements does Kant present something of an apology for his method in a manner that we can tie together with the whole of his thought:

> Nevertheless, just as we need a passage [*Überschritt*] from the metaphysic of nature to physics—a transition which has its own special rules—so we rightly demand something similar from the metaphysic of morals: a transition which by applying the pure principles of duty to cases of experience, would *schematize* these principles and present them as ready for moral-practical use. (*Akad.* §45, 468)

---

chapter to Kant's views on sex, and reaches a surprisingly sympathetic conclusion. He concedes at the outset that "Sexual morality is probably the very last topic on which a sensible person would want to defend Kant's views" (224). Kant does believe that since sexual pleasure involves degrading another person, it also degrades the one who feels it. On one hand, this view is difficult to grant. However, Wood argues, given that some measure of unsociable sociability or radical evil works against our rational nature and looks to subjugate others (especially women by men) in using them as a means, "there is at least a partial truth, however twisted it may be" (226).

Such transitions, Kant writes, do not belong to the pure part of the system. They can only be appended to (*angehängt werden können*) the system.[9]

Thus, matters culminate in what must be considered as a very odd amalgam containing inferences of varying kinds and strengths. The pure part of the system consists of synthetic a priori laws, that is, laws put together by imagination as ruled by reason: the moral law, the maxims that this law governs, and the typic. Taken together, this distances moral reasoning from both empiricism (there is no tie to pure intuition as the condition for empirical intuition) and mysticism (there is no tie to a non-sensuous intuition). However, the application of the pure part leaves a gap that is filled by what Kant calls *schematization*. This schematization must involve *empirical* schemata. This is the ultimate reason why moral matters seem impossible to settle except in very general terms.

The example of an empirical schema in the *Critique of Pure Reason* is that of the concept "dog," which Kant characterizes as a rule:

---

9 For an example of recent work in the tradition of Anglo-American Kant scholarship, see the much-discussed *Moral Literacy* by Barbara Herman (Harvard University Press: Cambridge, 2007]) in which she builds upon her preferred Kantian foundation to fashion a positive moral account that reaches out from Kant's "formalism" to other realms. Volume 16, Number 1 (2011) of *Kantian Review* is devoted almost entirely to a discussion of her book, which takes the path of "middle theory." Middle theory takes into account not only the individual moral subject, but also the "temporal aspect," of moral phenomena, their "dynamic aspect" (20 – 21) that includes society and its institutions. In "Herman on Moral Literacy," Stephen Engstrom finds reason to praise Herman's middle theory as "a practical understanding of one's socially constituted moral environment" (24), but wonders whether it is already included in our general moral nature or and also wonders whether it is a discernible skill (and what kind of skill it might be) in difficult cases. In "Will, Obligatory Ends, and the Completion of Practical Reason: Comments on Barbara Herman's *Moral Literacy*, Andrews Reath seems to wish that Kant provided a less rigid view that allowed for "an intermediate realm of value between the moral and the agreeable," (14) and supposes that such a realm might be what Herman offers. In "'Letting the Phenomena in'": On How Herman's Kantianism Does and Does Not Answer the Empty Formalism Critique" Sally Sedgwick, has worries from both sides. She hoped for "a more contentful account of the moral law" (43) on one hand, and on the other she was concerned that Herman's view threatened to abandon Kant's formalist commitments. In her response, Herman defends her claim that both a strong rationalism and a strong pluralism "one step down," (55) in which the categorical imperative unites with the obligatory ends so as to show us the form of good willing for us, can be successfully combined. Such combination allows for a plurality of moral thought since the specificity of ends might vary. The problem I have with all of this and with virtually all Anglo-American Kant scholarship is that in its abandonment of imagination, it also abandons both the possibility of principle-formation and its limits. Or in other words, the act of combination or putting-together, i.e., synthesis, is forgotten.

according to which my imagination can delineate the figure of a four-footed animal in a general manner, without limitation to any determinate figure such as experience, or any possible image that I can represent *in concreto*, actually presents. (A141, B180)

Since the schema in question is empirical, we may certainly err in its application. We may rather easily confuse a dog with a wolf at some distance, with certain cats, and with other "four-footed animals." We usually do not, but we certainly might.

How much greater is the room for error in the case of an empirical schema of a moral act! It is clear that there can be no schema of "moral" or "good." There can, however, be a schema of particular duties. Following Kant in his treatment of several perfect duties, there are indeed schemata that can effect the ruling of the proscription against suicide, but even these are vague. For example, under certain circumstances it *may* be permissible to commit what Kant calls self-murder: the suicide order by Nero of Seneca might be one such case. Even the death of a criminal can be ennobled somewhat by the "resoluteness with which he dies" (*Akad.* 436). I hasten to add that the vagueness of these empirical schemata accounts for some of Kant's conclusions that seem so bizarre by our standards, for example his anti-vaccination stance and his enthusiastic support for the death penalty. What differentiates Kant's position from most people who share his *particular* views is that their contemporary advocates have no moral theory at all, a shortcoming that is characterized by a mixture of resentment, bloodlust, and (occasionally) misinformation.

5 Conclusion

I shall begin to write my conclusion as a series of propositions that are rooted firmly in the Kantian text and have been treated above:
(1) Synthesis is the function of imagination.
(2) Synthetic judgments, one and all, include an act of imagination.
(3) The schemata are procedures of imagination that unite the Pure Concepts of Understanding with Pure Intuition.
(4) The moral law, its forms, and the maxims that flow from it, are one and all synthetic a priori judgments and therefore include imagination.
(5) The typic of the moral law is said to exclude imagination. Let this be stipulated for the reasons given by Kant.
(6) The typic enforces a very great gap between moral principles and their application in experience.

This gap is closed by *schematizing* these principles and presenting them as ready for moral-practical use.

(7) Imagination in its function as providing *empirical* schemata attempts to close this gap; however, it cannot do so in the matter of the pure schemata in theoretical philosophy.

(8) Therefore, moral arguments concerning particular issues can often remain contested, although the principles themselves cannot.

To some, this may appear to recollect W.C. Fields' deathbed Bible study. When visited by a friend who was appalled by atheist Fields' hypocrisy and who asked what on earth his friend was doing reading the Bible, Fields replied: "Looking for loopholes." I hasten to add that there are no loopholes in Kant's moral philosophy, only particular matters that must remain questionable. All moral actions must be done from duty, that is, in the spirit of duty, period. That is all that this moral philosophy, and perhaps any, can ask.

Fernando Costa Mattos

The Postulates of Pure Practical Reason

A Possible Place for Imagination in Kant's Moral Philosophy?

It is our purpose, in this essay, to discuss the role played by the faculty of imagination in Kant's practical philosophy, taking the postulates of pure practical reason as a special case in which we may notice such a role. But first we discuss the different interpretations of imagination in Kant's general philosophy, using those of Martin Heidegger and Hannah Arendt as two examples that, although very different from each other, recognize imagination as a central theme. With these – among other – interpretations as a point of departure, we then propose a reading of sections of the Critique of Practical Reason that present the postulates and try to show the role imagination may be playing (although unmentioned) in the creation of these postulates.

1 The Question of Imagination in Kant: between Heidegger and Hannah Arendt

It is common knowledge that the place occupied by imagination (*Einbildungskraft*) in the Kantian system is among the most controversial within scholarly debate. Also well-known – and no less polemic – is Martin Heidegger's stance, according to which imagination, being as it is the common root of the two major stems of human cognition, understanding and sensibility,[1] constitutes through schematism the very foundation of thought. To quote the words of one of his main works on Kant, *Kant and the Problem of Metaphysics,* of 1929:

> this pure schematism, which is grounded in the transcendental power of imagination, constitutes precisely the original Being of the understanding, the "I think substance," etc. As representing which forms spontaneously, the apparent achievement of the pure under-

---

1 Heidegger takes the notable passage from the "Introduction" to the *Critique of Pure Reason* as his basis: "…there are two stems of human cognition, which may perhaps arise from a common but to us unknown root, namely sensibility and understanding, through the first of which objects are given to us, but through the second of which they are thought" (KrV, A 16/ B 29). There is reasonable consensus among Kant scholars regarding the idea that imagination is indeed this common root of the two stems mentioned in the passage. On this particular question, there is a noteworthy contribution in Dieter Heinrich's article "On the Unity of Subjectivity" (in: Henrich, D. (1994): *The Unity of Reason*. Cambridge: Harvard University Press.

standing in the thinking of the unities is a pure basic act of the transcendental power of imagination.[2]

Not by chance, Heidegger draws mainly upon the first edition of the *Critique of Pure Reason* (1781) to support his interpretation, since this edition, in particular the version of the "Transcendental Deduction of the Pure Concepts of the Understanding" contained within it, treats imagination with much more emphasis than it would be in the second edition (1787) of the work. It is also well established that Heidegger tried to offer an explanation for this "regression" in Kant's work by pointing out the fact that the place occupied by imagination had not changed in the system of our cognitive faculties: the change in the handling of imagination would be the consequence of the choice to leave such an obscure theme further in the background, in order to concentrate first and foremost on the matters fundamental to the objective understanding of the world (which, in Heideggerian terms, constitutes the ontic understanding of reality— as opposed to the ontological— which is responsible for the handling of the subjective constitution of the human *Dasein*).

Be that as it may, the fact is that the Heideggerian interpretation ascribes an extremely privileged status to imagination. In contradiction to the majority of perspectives on the matter thus far, and especially against those of the neo-Kantian school of Marburg – whom Heidegger had faced off with in a celebrated gathering in Davos (one of the more relevant aspects that led to the development of *Kant and the Problem of Metaphysics*)[3] –, his own view supported the understanding of Kant's view of subjectivity as "laying the ground for metaphysics (*Grundlegung der Metaphysik*)", according to which imagination, our faculty responsible for synthesizing representations of time, would set the boundaries for the possibilities constitutive of our own limited being in the world, i. e. of our *Dasein*.

While it is not our purpose here to consider the details of Heidegger's interpretation of Kant,[4] we must still highlight briefly the function ascribed to imagination within this "fundamental ontology", as Heidegger called it, which Kant had established with the *Critique of Pure Reason*. It must be pointed out, that

2 Heidegger, M. (1997): *Kant and the Problem of Metaphysics*. Transl. Richard Taft. Indianapolis: Indiana University Press, p. 106.

3 In the preface to its first edition, Heidegger himself refers to the lectures held at Davos as a form of preparation for the book. Cf. Heidegger, *op. cit.*, p. xix.

4 For a quite detailed discussion of this interpretation, there is a recent and ambitious work worthy of attention: Rebernik, P. (2006): *Heidegger interprete di Kant. Finitezza e fondazione della metafisica*. Pisa: Edizioni ETS.

the argument does not consist of regarding imagination as the main faculty responsible for scientific knowledge –that is a specific function of understanding alongside sensibility. What matters here is demonstrating that, imagination being the common stem, the mediation between the faculties of understanding, within imagination itself is where we should look for the key to the essential (ontological) understanding of the structures of *our* world, i. e. of the world as it is constituted from our own rational, human and finite perspective (within which scientific knowledge is *just one*, among several possibilities).

Heidegger is not the only interpreter of Kant who places imagination in a prominent position, but others who have done so followed a quite distinct exegetical direction. A good example thereof, which has been a reference since the 1970 s, is the interpretation put forth by Hannah Arendt, who bases herself not so much on the first, but specifically on the third *Critique* (the *Critique of Judgment*). She is far from attempting to put into question the fundamental basis of critical philosophy, or indeed its place in the history of western metaphysics, but, rather, seems to be inspired by some key concepts of the *Critique of Judgment* – notably that of *sensus communis* and reflective judgment, in both of which imagination plays a foundational role. In doing so, Arendt rethinks Kant's political philosophy or, perhaps more decidedly, rethinks *politics* itself in light of Kant's philosophy. Let's take, for example, the following passage from her *Lectures on Kant's Political Philosophy*, well-known in the English speaking world:

> Critical thinking is possible only where the standpoints of all others are open to inspection. Hence, critical thinking, while still a solitary business, does not cut itself off from "all others." To be sure, it still goes on in isolation, but by the force of imagination it makes the others present and thus moves in a space that is potentially public, open to all sides; in other words, it adopts the position of Kant's world citizen. To think with an enlarged mentality means that one trains one's imagination to go visiting.[5]

As we can see, imagination plays an essential role therein. It is however an entirely different role than the one it played in Heidegger's interpretation: departing from the basis of theoretical knowledge as already established in the *Critique of Pure Reason*, the task at hand is to explore – be it in the third *Critique*, be it in the "minor" political works – the possible consequences of our reflexive mode of thought in areas which may not be subject to solutions elaborated purely within the epistemological key given in the first *Critique*. If we need to "leave" our own point of view (a point of view which finds its constitutive character in determi-

---

5 Arendt, H. (1992): *Lectures on Kant's Political Philosophy*. Chicago: The University of Chicago Press p. 43.

nant judgments)[6] so that we, by means of placing ourselves in another person's shoes, may consider issues of a collective nature (a process which, according to Arendt, would constitute a reflexive political judgment), then imagination must come into play with one of its more, let's say, "traditional" functions – shifting us mentally to an imaginary position, distinct from the one we effectively occupy.

Naturally, this process articulates itself directly in the manner in which imagination operates in the reflexive aesthetical judgment, itself a theme Arendt explores in depth – as is the case with many other scholars who followed a similar interpretive strategy.[7] In fact, the idea that a certain type of judgment constitutes itself through the free interplay of understanding and imagination, as the *Critique of Judgment* argues, is very inspirational regarding some particular structures of thought. Though lacking objectivity *per se*, it doesn't mean they are not based on a kind of derived "objectivity", comprehensible in terms of an inter-subjective entity which would be implicit to a degree and negotiable to a degree (in a political sense). In essence, the perspective put forth by Arendt consists of the unfolding of this discovery – concerning, at first, solely aesthetic judgment – to other realms of thought and, in particular, to politics.

The most obvious pathway, then, to those who are concerned with the question of the role of imagination in Kant's practical philosophy, would be to follow in the footsteps of Hannah Arendt and, retrospectively[8], to bring elements of the

---

6 It could be argued that inter-subjectivity, and even the ability to place ourselves in another person's point of view, are essential for any type of judgment, including of a determining nature. Onora O'Neill, for example, argues something similar to this. Though, to even admit this possibility, it would be necessary to recognize a substantial difference in inter-subjectivity's degree of importance – and of imagination as well – for two types of judgment (unless we were to go back to the Heideggerian stance, which is not the case either in O'Neill's or in Arendt's work). Cf. O'Neill (1989): O. *Constructions of Reason. Explorations of Kant's Practical Philosophy*. Cambridge: Cambridge University Press.

7 Following the example set by Daniel T. Peres, a Brazilian Kant scholar who has dedicated himself to the theme of imagination within practical philosophy, it would be interesting to mention the names of Cornelius Castoriadis, Alain Renaut and François Lyotard, besides the recent work of Jane Kneller, which has already become a reference in the studies on imagination in Kant's work: Kneller, J. (2007): *Kant and the Power of Imagination*. Cambridge: Cambridge University Press. Cf. Peres, D. "Imaginação e razão prática". In: *Analytica*. Rio de Janeiro, vol.12, n.1, 2008, p. 99–130. There is a shorter version of the article in English: Peres, D. "Imagination and Practical Reason". In: *Kant e-Prints*. Campinas, Série 2, v. 3, n. 2, p. 293–296, jul.-dez., 2008 (link: http://www.cle.unicamp.br/kant-e-prints/index_arquivos/kant-vol3-n2.htm).

8 As Peres puts it, "any interpretation of the relationship between imagination and practical reason in Kant, even if running the risks inherent in a retrospective analysis (a possible result of shedding light on the whole of the critical works from a perspective which presents itself only in 1790), any such interpretation must set out from the *Critique of Judgment*. It is better to do a

Critique of Judgment, or even from the shorter works on political philosophy[9], to bear on the inner workings of the *Critique of Practical Reason* itself. This attempt may mitigate the resilience of the well-known statement, present in such work, according to which imagination would have no function whatsoever in practical judgment.[10] There is something at once strikingly simple and promising in the argument: taking the notion of "exemplary validity" in the manner it is descri-bed by Kant in § 76 of the *Critique of Judgment* (regarding the question of the tel-eological reflexive judgment) as a basis, the task would be to find in human ac-tions examples illustrative of general imperatives truly compatible with moral law.[11]

Though this naturally does not affect the foundations of the categorical im-perative, which are strictly rational (and, therefore, quite independent of imagi-nation's activity), it is nonetheless an interesting key through which one can con-sider the real *application* of this imperative, be it from an individual point of view, be it from humanity's or a historical point of view.[12] If we think about the *Groundwork of the Metaphysics of Morals* itself, in particular the examples drawn from popular moral wisdom in the book's first sections, it would be pos-sible to detect such a procedure – the "political judgment", as Arendt sees it – in full effect.

Kant himself tirelessly stated that examples cannot serve as a basis for mor-ality, and can only indicate (*de facto*) to the philosopher that interprets them what their conditions of possibility would be (*de juris*). But it is not a question here, as we stated before, of basing morality on examples, but, instead, to take them as symbols of universal rules without which the individual could

---

retrospective analysis that is based on the works themselves, than to overreach and refer to a profound perspective that has never been thought before, to which inaccuracy is the only possible outcome". Peres, D., *op. cit.*, p. 103. It's clear that Peres refers to the Heidegger-inspired interpretations, particularly that of Bernard Freydberg, to which we'll come back. Freydberg, B. (2005): *Imagination in Kant's 'Critique of Practical Reason'.* Indianapolis: Indiana University Press.

9 Works such as, for instance, *Idea of a Universal History from a Cosmopolitan Point of View, What is Enlightenment?, What does it Mean to Orient Oneself in Thinking.*

10 KpV, Ak.V, 69: "...the moral law has no cognitive faculty other than the understanding (not the imagination) by means of which it can be applied to objects of nature, and what the understanding can put under an idea of reason is not a *schema* of sensibility but a law, such a law, however, as can be presented *in concreto* in objects of the senses and hence a law of nature, though only as to its form".

11 Arendt, H. *op. cit.*, p. 84.

12 A point of view which is adopted by Kant in works such as *Idea of a Universal History From a Cosmopolitan Point of View* and *Towards Perpetual Peace.*

not consider himself a moral being. This would also be valid *a fortiori* to the realm of political philosophy, in which events like the French Revolution, as one can see in Kant's own reception of the event, play the role of a sign that indicates the possibility of moral progress of the species.

We also see this in the case of the postulates of practical reason, another important reference in Kant's moral philosophy in regard to the *application* of moral law upon the world. According to this point of view, these postulates too could be read as elements of a politically engaged view of history – forged in the context of political judgment – aimed quite simply at reinforcing the possibilities of action set by moral law as a categorical imperative of right. God and the immortality of the soul would not be, therefore, much more than auxiliary hypotheses to reinforce the conviction of the progress of human institutions, the design of which would be taken on by imagination as an active faculty in the establishment of political judgment (satisfying the desire of authors like Castoriadis, who complained of a certain dearth of all things imaginative in Kant's political philosophy).[13]

But the truth of the matter is that such interpretations, by radicalizing the political element of Kantian philosophy, tend to pay little attention to the issue of the postulates, above all with respect to their semantic proximity to the metaphysical tradition. It is a known fact that, ever since the time of Kant, there have been repeated accusations that charged the philosopher with "taking a step back" with respect to the progressive aspects of the *Critique of Pure Reason*, reincorporating old metaphysical ideas under a new guise (so-called "practical knowledge") without essentially altering them.[14] Insisting on the theme of the postulates itself, from this point of view, would be counterproductive, for a long discussion would ensue only to explain that, despite appearances, they have no relation with the Christian metaphysical tradition (a relation that, in

---

13 Cf. Castoriadis, C. (1991): "The greek polis and the creation of democracy". In: *Philosophy, Politics, Autonomy. Essays on Political Philosophy*. New York, Oxford: Oxford University Press, p. 81–123. Quoted in Peres, D., op. cit., p. 101.

14 I've written a brief essay on the theme of "practical knowledge" in an article published in the annals of the X International Kant Congress of 2005. Cf. Mattos, F. (2008): "Kant's practical knowledge as a result of the connection between speculative metaphysics and rational faith." In: Terra, R., Ruffing, M. et. al. *Recht und Frieden in der Philosophie Kants. Akten des X. Internationalen Kant-Kongress*. Berlin, New York: Walter de Gruyter, vol. 3, p. 259–268. The essay is based on my Master's dissertation at the University of São Paulo, which had its public defense in 2001. Cf. Mattos, F. (2001): *Conhecimento prático e metafísica especulativa em Kant* ("Practical Knowledge and Speculative Metaphysics in Kant"). Dissertação de Mestrado, 117 pages. São Paulo: Department of Philosophy.

general, is seen as detrimental to the perspective of political progress of human-ity).[15]

However, the question concerning this issue that perhaps should be asked is precisely the one regarding how big a fracture there would really be in Kant with respect to his world view that, according to himself, would be more adequate to the needs of reason (*Bedürfnisse der Vernunft*) when one attempts to consider the future through moral law. Gottfried Martin, for one, suggests that there is a sig-nificant degree of continuity between Kantian thought and the scholastic tradi-tion, which would be most visible in the new "determinants" of being (of man, of the world and of God) within the Kantian conceptual framework. The novel as-pect therefore would consist less in hollowing out the meaning of the old meta-physical notions and more in the particular approach taken by the philosopher, an approach that is rigorously humanistic, i. e. a point of view that sets out from that which is foundational with respect to our way of seeing the world (from the subjective forms of sensibility – space and time – to the ideas of reason, in ad-dition to the categories of understanding that make it possible for us to think of the object we perceive intuitively).

To consider God or the soul from this point of view means serving a need we have, a need that stems from the incompleteness of our empirical knowledge of reality (which we can perhaps trace back to our essential finiteness). The need stems from reason's "natural vocation" to go beyond the limits of sensible expe-rience. But this "going beyond" would have no significance (*Bedeutung*) if it didn't involve that capability of synthesis with which imagination articulates dif-ferent notions (giving a predicate to a subject, for example). It is inevitable in other words, that we *imagine* God and the soul when we think about them, and that is something we can only accomplish by *making these notions sensible* (*versinnlichen*), i. e. by attributing properties to them that we know from sensible reality. Furthermore, if Kant's efforts in the "Dialectic" of the first *Critique* go in the direction of avoiding an error that would be detrimental to the theoretical

---

15 We cannot forget, however, the valuable contributions that have deemed it necessary to insist on this argument. A notable case is that of Lewis W. Beck, who in his classical commentary to the *Critique of Practical Reason* sets out to demonstrate the inexistence of any connection whatsoever between the doctrine of the postulates and the foundations of morality, which would be strictly rational. Beck seems even to be upset that Kant had given attention to these old themes. Cf. Beck, L. (1984): *A Commentary on Kant's Critique of Practical Reason*. Chicago: The Chicago University Press(2nd.ed.), p. 263. Another important name to be mentioned here, in the context of continental Europe, is that of Gérard Lebrun, who, in his *Kant et la Fin de la Méta-physique*, also seeks to show that these "old concepts" – such as God, the soul – actually don't refer to any object in particular, rather constituting a "radioscopy of meanings". Cf. Lebrun, G. (1970): *Kant et la Fin de la Métaphysique*. Paris: Armand Colin, p. 263.

knowledge of the world, he would adopt a different stance regarding them in the second *Critique*, recognizing the importance they possess, even as mere hypotheses (which, we must insist, mean nothing without their becoming sensible through imagination), as a means to satisfy another sort of need stemming from reason, now practical instead of theoretical.

It is quite evident – any reader familiar with Kantian terminology would be quick to protest – that our terms do not strictly follow the lessons of Kant: besides the previously highlighted fact that Kant expressively deprives imagination of any function in practical judgment, he also doesn't seem to grant any space to this faculty in the "Antinomies" of the first *Critique*, much less in the "Dialectic" of the second one. That the enunciation of the postulates could be taken as a basis to "determine" (in spite of the obligatory quotes there) man's or God's being, as Martin would have, is something that seems at first sight to contradict not only the written word of Kantian philosophy but its deep-seated spirit as well. Nevertheless, we shall insist on this particular argument, clearly Heideggerian in its inspiration, to verify[16] if the doctrine of the postulates wouldn't be better served when we proceed in such a manner. We shall try to do so without going against the essential components of Kantian philosophy.

2 Searching for the Footprints of Imagination in the "Dialectic of Pure Practical Reason"

We will begin with Kant's own remarks. After opening the "Dialectic of Pure Practical Reason" with a short – and quite generic – chapter entitled "On the Dialectic of Pure Reason in General", Kant goes on to characterize the dialectic of practical reason as a means to determine the "concept of the highest good" and, in the lines that follow, he starts to clarify, in a very precise manner, the core of the issue that constitutes "the antinomy of practical reason":

> In the highest good which is practical for us, that is, to be made real through our will, virtue and happiness are thought as necessarily combined, so that the one cannot be assumed by pure practical reason without the other also belonging to it. Now, this combination is (like every other) either *analytic* or *synthetic*. Since, as has already been showed, the given combination cannot be analytic, it must be thought synthetically and, indeed, as the connection of cause and effect, because it concerns a practical good, that is, one that is possible through action.[17]

---

16 In line with Bernard Freydberg, whom we already mentioned.
17 KpV, Ak.V, 113.

The first notable feature of this passage, which marks the beginning of "the antinomy of practical reason", is the reference to the synthetical element which plays – as it must – an important role in the bond between virtue and happiness, and which also constitutes the antinomy's central issue. Well! If we recall that imagination, as a "blind but indispensable function of the soul, without which we would have no knowledge whatsoever", is the faculty responsible for the "synthesis in general", then this would constitute a first argument in favor of recognizing its presence here, even though we are "scarcely ever conscious" of it.[18] Despite the fact that it is an indirect presence, so to say, it would serve at one time as primary and a more general evidence – here and in any other synthetical judgments in the *Critique of Practical Reason* – for Freydberg's acute argument according to which "imagination runs through the *Critique of Practical Reason* as its mostly concealed and silent but nevertheless guiding thread".[19]

It is evident that imagination would not be able to operate on the surface, since it is connected to a set of problems that manifest themselves only at the conceptual level. At the conceptual level, an apparent contradiction between the realms of the theoretical (the empirical comprehension of the world, a sphere where man is inclined to follow his impulses in the search for happiness) and the practical (moral law through which man shows himself capable of action in accordance to a causality distinct from nature) establishes itself. But the solution to this conundrum, which would, in principle, require an impossible synthetical connection, must involve imagination in some manner. Let's refer back to the remarks of Kant himself, who, after arguing against the possibility of a linear implication between two different instances – be it from virtue to happiness, or the other way around[20] – points out an apparent lack of solution on the horizon:

---

18 We paraphrase the well-know passage in B 103: "Synthesis in general (*überhaupt*) as we shall hereafter see, is the result of the power of imagination, a blind but indispensable function of the soul, without which we would have no knowledge whatsoever, but of which we are scarcely ever conscious" (KrV, A 78, B 103).

19 Freydberg, *op. cit.*, p. 3.

20 After the passage we've just quoted, Kant goes on to say that : "The first is *absolutely impossible* because (as was proved in the Analytic) maxims that put the determining ground of the will in the desire for one's happiness are not moral at all and can be the ground of. But the second is *also impossible* because any practical connection of causes and effects in the world, as a result of the determination of the will, does not depend upon the moral dispositions of the will but upon the knowledge of the laws of nature and the physical ability to use them for one's purposes; consequently, no necessary connection of happiness with virtue in the world, adequate to the highest good, can be expected from the most meticulous observance of moral laws" (KpV, Ak.V, 113).

> Now, since the promotion of the highest good, which contains this connection in its con-
> cept, is an a priori necessary object of our will and inseparably bound up with the
> moral law, the impossibility of the first must also prove the falsity of the second. If, there-
> fore, the highest good is impossible in accordance with practical rules, then the moral law,
> which commands us to promote it, must be fantastic and directed to empty imaginary ends
> and must therefore in itself be false.[21]

Although, in principle, the impossibility of the highest good should not compromise the basis of morality[22], as we have claimed so far, Kant here seems to indicate that, if it is indeed impossible, then the latter (morality) would prove to be a
chimera, compromising the entire critical project as a whole (if we are to take
Kant's many claims in this respect, included the "Dialectic of practical reason").
An interesting aspect of this argument is the expression chosen by Kant to characterize the fate of moral law if this were the case (i. e., if the highest goods were
in fact impossible): it would only serve "empty imaginary ends" (*leere eingebildete Zwecke*). The fact that Kant characterizes moral ends as "imaginary" or,
rather, "imagined" (perhaps a more precise translation), does not mean that
moral ends can't be envisioned: what they can be is empty, as they well should
be if the highest goods were impossible in the world. If moral ends were imagined (as they must be, if they are synthetical) as well as *meaningful* – which is to
be shown in the "critical solution of the antinomy" – then it follows that they
must represent the combined operation of our mental faculties in the intellectu-
al-imaginative organization of the world that, at the bottom, is the most general
of the objects of Kantian critical philosophy. It isn't a matter of chance that Kant
draws a parallel to the antinomy of speculative reason as his starting point in the
development of a "critical solution of the antinomy of practical reason", indicating the need to refer once more to the distinction between appearances and the
thing in itself.

> In the antinomy of pure speculative reason there is a similar conflict between natural ne-
> cessity and freedom in the causality of events in the world. It was resolved by showing that
> there is no true conflict *if the events and even the world in which they occur are regarded
> (and they should also be so regarded) merely as appearances*; for, one and the same acting
> being as *appearance* (even to his own inner sense) has a causality in the world of sense that
> always conforms to the mechanism of nature, but with respect to the same event, *insofar as
> the acting person regards himself at the same time as a noumenon* (as pure intelligence, in

---

21 KpV, Ak.V, 114.
22 This is a fundamental point for Beck: "We must not be deceived, as I believe Kant was, into
thinking its possibility [of the highest good] is directly necessary to morality or that we have a
moral duty to promote it, distinct from our duty as determined by the form and not the by the
content or object of the moral law". Beck, L., *op. cit.*, p. 245.

his existence that cannot be temporally determined), he can contain a determining ground of that causality in accordance with laws of nature which is itself free from all laws of nature.[23]

The explanation given here is similar to the one Kant provided in the third section of "Groundwork of the metaphysics of morals". In both cases the "vicious cycle" was solved with the argument that the apparent contradiction of the antinomy disappears when we realize that, if we do not take the sensible world as reality *itself* in absolute terms (since it is but *a phenomenal* reality), we may consider ourselves not only as sensible beings, determined by the causality of nature, but also *noumenal*, as beings capable of acting according to a form of causality that is "free from all laws of nature". And the aspect we must highlight in the present context is the movement of "thought" by which we place ourselves, back and forth, within the perspective of the phenomena and within the perspective of the thing itself. This process calls to mind the movement of imagination, held dear by Hannah Arendt, through which we assume the standpoint of someone else. This leads us to question whether in this instance the role of imagination, even if implicit, wouldn't have a higher degree of importance.

This is precisely what Bernard Freydberg suggests, recalling that reason, without the support of imagination, wouldn't be able to perform syntheses or even create images which are essential for us to truly reflect upon the nature of our own being. This is the case both when we do it from a sensible standpoint, in which we only comprehend appearances thanks to the syntheses operated by imagination in the sensible world, and when we do it from an intelligible standpoint, in which one is only able to think of oneself as a thing in itself when we *imagine*, by means of an analogy with the sensible world, an intelligible world that goes as far as moral law needs it to achieve its aims. This is how, starting from this first "stepping stone" into the non-sensible – which is, at the bottom, the concept of freedom actively operating in our minds – we begin to form (*bilden, einbilden*) a picture of the intelligible world which, though "imaginary" or "imagined" (*eingebildet*), is no longer empty because we fill it with "practical elements", as Kant argues in a well-known passage of the second preface to the *Critique of Pure Reason* of 1787 (where Kant, having already written the *Groundwork*, had his sights set on the *Critique of Practical Reason*). In other words, we fill it with syntheses of the imagination that are not just allowed, but rather demanded by our morality especially when it comes to putting the categorical imperative into effect.

---

23 KpV, Ak.V, 114 (the emphasis is mine).

That is the reason why freedom, even if there is no section in the book so entitled (like the immortality of the soul and the existence of God), constitutes the first postulate of practical reason[24]: it is freedom, the key that unveils the possibility to assume other standpoints, which provides the solution to the antinomy of pure practical reason. It does so by making natural necessity, which belongs to the sensible, phenomenal, world, and the feasibility of the highest good commanded by moral law (now seen as a foundation of our intelligible being) compatible. And it is precisely this compatibility, made possible by freedom, which will give way to two other postulates (which correspond to what is out of reach in the world, but is nevertheless necessary to conserve our expectation of fulfillment of moral law) described in the final remarks of the "critical solution of the antinomy":

> since the possibility of such a connection of the conditioned with its condition belongs wholly to the supersensible relation of things and cannot be given in accordance with the laws of the sensible world, although the practical results of this idea – namely actions that aim at realizing the highest good – belong to the sensible world, we shall try to set forth the grounds of that possibility, first with respect to what is *immediately* within our power and then, secondly, in *that which is not in our power but which reason presents to us, as the supplement to our inability, for the possibility of the highest good* (which is necessary in accordance with practical principles).[25]

The terms utilized here by Kant seem to reinforce the connection, pointed out in the theoretical context by Heidegger and explored in the practical by Freydberg, between the manner in which reason thinks out (imagines) the supersensible, on the one hand, and finiteness, on the other, the latter being its constitutive element as it is of human *Dasein* as a whole.

It is, in fact, a matter of "supplementing our inability" when we speak of putting the highest good into effect in the world, a goal that we ourselves have set from an intelligible standpoint, through moral law. In the same way

---

24 When he comments on the postulates as a whole, in the section entitled "Of the postulates of Pure Practical Reason in general", Kant summarizes them with the following words: "These postulates are those of *immortality*, of *freedom* considered positively (as the causality of a being insofar as it belongs to the intelligible world), and of the *existence of God*. The *first* flows from the practically necessary condition of a duration befitting the complete fulfillment of the moral law; the *second* from the necessary presupposition of independence from the sensible world and of the capacity to determine one's will by the law of an intelligible world, that is, the law of freedom; the *third* from the necessity of the condition for such an intelligible world to be the highest good, through the presupposition of the highest independent good, that is, of the existence of God." (KpV, Ak.V, 132)

25 KpV, Ak.V, 119 (the emphasis is mine).

that our understanding, in a theoretical context, stamped necessity and universality upon the natural world, two things that, in our elementary finitude (the boundaries of which are set by sensibility), we were not able to ensure (something visible in the manner in which Hume put forth the problem of causality), reason's role is to ensure that the world (in general) may have a moral direction, which we, in our finitude, are not able to provide. The ideas of the soul and God, which appeared as regulatory hypotheses in the theoretical context, receive a predicative complementation (the soul *is immortal* and God *is the moral source of the world*)[26] and convert themselves into postulates whose "efficacy", if we are to believe in them, is not an inch smaller than that of theoretical knowledge.[27]

This is a plausible reason for the fact that Kant introduces a section on the primacy of practice between the "critical solution of the antinomy" and the sections that specifically handle the postulates of the immortality of the soul and of the existence of God: it seems it is necessary to reinforce the existing hierarchy between practical reason – which gives us meaning and purpose in life – and its theoretical counterpart, whose activity could not be considered an end in itself, for it must submit to practical reason when we have our entire existence in view (not only its cognitive dimension). By making this evident, Kant sets the stage to establish, in the form of the postulates, those synthetic connections which, being founded on freedom, shall reinforce the bond between virtue and happiness as the endless progress of the human species:

> This endless progress is, however, possible only on the presupposition of the *existence* and personality of the same rational being continuing *endlessly* (which is called the immortality of the soul). Hence the highest good is practically possible only on the presupposition of the immortality of the soul, so that this, as *inseparably connected with the moral law*, is a *postulate* of pure practical reason (by which I understand a *theoretical* proposition, though one not demonstrable as such, insofar as it is *attached inseparably to an a priori unconditionally valid practical law*).[28]

---

26 Cf. Mattos, F. "Kant's practical knowledge as a result of the connection between speculative metaphysics and rational faith", p. 263 ff.

27 See, for example, *What does it mean to orient oneself in thinking*, Ak.VIII, p. 141: "...*rational faith*, which rests on a need of reason's use with a *practical* intent, could be called a *postulate* of reason – not as if it were an insight which did justice to all the logical demands for certainty, but because this holding true (if only the person is morally good) is *not inferior in degree to knowing*" (the last emphasis is mine).

28 KpV, Ak.V, 122 (the last emphasis is mine).

In opposition to those who set out to interpret the mortality of the soul as meaning the perpetual existence of the human species, this passage shows that it is actually an assumption relative to "the same rational being" (*dasselbe vernünftige Wesen*). Moreover, it indicates that its connection to moral law is "inseparable" (*unzertrennlich*). It is evident that this assumption cannot strictly constitute knowledge as elaborated in the first *Critique* (this is an achievement, so to say, that critical philosophy has never renounced), its nature being that of a theoretical proposition (therefore a synthetic one) and, in light of that, imagination having a role in it. What confers legitimacy to it – another aspect that the passage makes clear – is the fact that this connection with moral law, which makes it an "inseparable result" stemming not from an individual subjective arbitrariness (as would be the case in *leere eingebildete* products of imagination), but from "an unconditionally valid practical law", the same law that

> must also lead to the possibility of the second element of the highest good, namely, *happiness* proportioned to that morality, and must do so as disinterestedly as before, solely from impartial reason; in other words, it must lead to the supposition of the existence of a cause adequate to this effect, that is, it must postulate *the existence of God* as belonging necessarily to the possibility of the highest good (which object of our is necessarily connected with the moral lawgiving of pure reason).[29]

The existence of God comes as a complement, in this sense, to the group of three postulates that, corresponding to the ideas with which speculative reason had sought to solve its own apparent contradictions (paralogisms, the antinomies and the ideal of completeness) make it possible to establish a "moral world view" (to use the derogatory terms of Hegel without any derogatory intentions of our own) in which freedom is seen as converging, in a hypothetical future (which makes putting it into effect an unending task) with an empirical reality with which a rational being was not able to, in principle, identify fully, for at heart he is no more than the sum of possibilities set by the strict conjunction of sensibility and understanding. Naturally, it's not the case of believing in a God overloaded with predicates, as religions in general have painted him: it is a question of conceiving only a "moral author of the world" that, besides granting a hypothetical completeness to our perception of the world (a perception limited by sensibility, i.e., by our finiteness), can also indicate a moral pathway to the world, i.e., point it in the direction of a progressive implementation (though an interminable one) of this moral law, which, according to our own "self-comprehension" (dissected by Kant in his analytic portrayal of human *Dasein*, as

---

29 KpV, Ak.V, 124.

Heidegger would say), is foundational to our being (a *factum* of reason, in the *Critique of Practical Reason*'s own terms).

The fact that this self-comprehension does not imply a broadening of our knowledge of the world is something Kant never ceases to point out. But he also insists on the synthetic element of unification (*Vereinigung*), which is present in this movement, through which our mind (by means of the imagination), having always moral law as its guiding thread, amplifies our "knowledge" "in a practical point of view" (*in praktischer Absicht*):

> But is our cognition really extended in this way by pure practical reason, and is what was *transcendent* for speculative reason *immanent* in practical reason? Certainly, but only *for practical purposes*. For we thereby cognize neither the nature of our souls, nor the intelligible world, nor the supreme being as to what they are in themselves, but have merely unified (*vereinigt*) the concepts of them in the *practical* concept *of the highest good* as the object of our will, and have done so altogether a priori through pure reason but only by means of the moral law and, moreover, only in reference to it, with respect to the object it commands.[30]

As was said previously, the major question incited by this "trick" of Kant's, with which he "revalidates", so to say, the classical concepts of metaphysics, God, freedom and the soul as the bases of the three branches of *metaphysica specialis*, is related to the extent in which he would, in doing so, compromise the integrity of his own critical philosophy. Well! If the primary aspect of metaphysics – ontology – concerns our mental faculties, this means it is, first and foremost, an ontology of human being: it is in this sense that Heidegger considers it a re-foundation of metaphysics in the analytics of human finiteness. It is precisely here where the radical nature of its critical stance would be most visible: overthrowing the bases of what used to be an ontology of the world, he places himself in the standpoint of human being (an "all too human" standpoint, if we think in the words of Nietzsche) to investigate the horizons of possibility offered to us as finite beings, placed in a temporal reality that, within us, manifests itself in the *power* of our imagination (and, in this respect, we must emphasize the *Kraft* in *Einbildungskraft*).

In this sense, the possibilities unveiled to our interpretation of the world by scientific knowledge are just a small part (that which corresponds to the conjunction between understanding and sensibility in a synthesis that articulates sensible data with intellectual concepts) within the grand sum of possibilities uncovered by Kant in our subjectivity. The value of this knowledge (*Wissen*) is, to Kant, in the fact that it is, as a mode of acceptance, both subjectively and ob-

---

30 KpV, Ak.V, 133.

jectively sufficient (*zureichend*), while rational faith (*vernünftiges Glauben*), a kind of acceptance that will tie itself to the postulates of practical reason, would only be enough subjectively. Well! If we remind ourselves that the raw material utilized by imagination is also sensible, everything it does, when connecting predicates to the ideas of reason, is to put aside the guiding thread of understanding (which, when tied to objects, would lead to a predominance of theory) and to adopt instead the thread of practical reason, which demands certain pathways to combine this material, suitable as it is to our self-comprehension as essentially moral beings (all this within a single ontology of human subjectivity). This other "pathway to combine", we insist, is not arbitrary in any way (subjective insufficiency, a characteristic of opinion, would indicate arbitrariness), for it is above all

> *a need from an absolutely necessary point of view* and justifies its presupposition not merely as a permitted hypothesis but as a postulate from a practical point of view; and, granted that the pure moral law inflexibly binds everyone as a command (not as a rule of prudence), the upright man may well say: I *will* that there be a God, that my existence in this world be also an existence in a pure world of the understanding beyond causal connections, and finally that my duration be endless; I stand by this, without paying attention to rationalizations, however little I may be able to answer them or to oppose them with others more plausible, and I will not let this faith be taken from me; for this is the only case in which my interest, because I *may* not give up anything of it, unavoidably determines my judgment.[31]

It wouldn't be an exaggeration at this point to say that (according to the interpretation we have been advancing), this is the way the *imagination* of the upright man functions, according to Kant. Though understanding, whose function is to apply concepts to sensible data, would never allow such conclusions to be reached, imagination, *guided by morality* – and by the interest this law awakens in us – establishes connections that lie beyond the sensible (or beyond the sensible as it is conceptualized by understanding, i.e., our empirical knowledge) which allow them to conserve hope that this moral interest, and acting in accordance to it, are not done in vain. This says very little about the world, as Kant tirelessly argues, but it says a great deal about ourselves, finite beings which become aware of the fact that we exist in this world, that we *are* in this world (*in-der-Welt-sind*) without having an adequate comprehension of it. And this perspective might clarify the often quoted passage of Kant, in the end of the *Critique of Practical Reason*, according to which "two things fill the mind with ever new and increasing admiration and reverence, the more often and

---

31 KpV, Ak.V, 143.

the more steadily one reflects on them: *the starry heavens above and the moral law within me"*.

3 Concluding Remark

The interpretation suggested here is certainly not the most orthodox. By insisting on the importance of imagination, we run the risk of falling prey to imagination ourselves. Freydberg himself, whose arguments strongly influenced our essay, seems to go a bit too far when he talks about a "primacy of imagination" in Kant's philosophy.[32] Supporting a slightly less ambitious perspective, we have made a point of insisting on the fact that, by venturing beyond the sensible, imagination (and with it thought, mind) operates under the command of moral law, therefore, under the predominance of practical reason – something about which Kant is particularly clear. Nevertheless, we have sought to show that imagination, being fundamental to thought in the theoretical realm, also must have its practical function recognized, particularly with respect to the postulates of practical reason: the immortality of the soul, freedom, and the existence of God. When we proceed in such a manner, Kantian moral philosophy effectively acquires less of a gray tone (as Freydberg argues),[33] but not to the point where it deviates from its rationalist and universalist core, or from its place within an ontology of finitude that, if properly understood, can offer a much wider array of possibilities to human being than a mere theory of knowledge (Cassirer) or a moral philosophy centered on politics alone (Hannah Arendt).

References

Arendt, H. (1992): *Lectures on Kant's Political Philosophy*. Chicago: The University of Chicago Press.

Beck, L. (1984): *A Commentary on Kant's Critique of Practical Reason*. Chicago: The University of Chicago Press.

Castoriadis, C. (1991): *Philosophy, Politics, Autonomy. Essays on Political Philosophy*. New York, Oxford: Oxford University Press.

---

32 Cf. Freydberg, *op. cit.*, p. 105: "But since imagination drives all synthesis and therefore all rational activity, the primacy of the practical is ultimately... the primacy of imagination".

33 Cf. Freydberg, p. 22: "...the exposure of imagination at the heart of Kant's practical philosophy may serve to make this philosophy—too often characterized as joyless and as at least partially unsuited to our nature—appear in a friendlier light".

Freydberg, B. (2005): *Imagination in Kant's 'Critique of Practical Reason'*. Indianapolis: Indiana University Press.

Heidegger, M. (1997): *Kant and the Problem of Metaphysics*. Transl. Richard Taft. Indianapolis: Indiana University Press.

Henrich, D. (1994): *The Unity of Reason*. Cambridge: Harvard University Press.

Kant, I. (1996a): "Critique of Practical Reason". Transl. Mary J. Gregor. In: Kant, I. Practical Philosophy. Cambridge: Cambridge University Press,. Kant, I. (1996b): "What does it mean to orient oneself in thinking?". Transl. Allen Wood. In: Kant, I. *Religion and Rational Theology*. Cambridge: Cambridge University Press.

Kant, I. (1998): *Critique of Pure Reason*. Transl. Paul Guyer, Allen Wood. Cambridge: Cambridge University Press.

Kneller, J. (2007): *Kant and the Power of Imagination*. Cambridge: Cambridge University Press.

Lebrun, G. (1970): *Kant et la Fin de la Métaphysique*. Paris: Armand Colin.

Mattos, F. (2008): "Kant's practical knowledge as a result of the connection between speculative metaphysics and rational faith." In: Terra, R., Ruffing, M. et. al. *Recht und Frieden in der Philosophie Kants. Akten des X. Internationalen Kant-Kongress*. Berlin, New York: Walter de Gruyter, vol. 3, p. 259–268.

O'Neill, O. (1989): *Constructions of Reason. Explorations of Kant's Practical Philosophy*. Cambridge: Cambridge University Press.

Peres, D. (2008): "Imaginação e razão prática". In: *Analytica*. Rio de Janeiro, vol.12, n.1, p. 99–130.

Rebernik, P. (2006): *Heidegger interprete di Kant. Finitezza e fondazione della metafisica*. Pisa: Edizioni ETS.

Jane Kneller
Imagining our World

Affinity and Hope in Kant's Theory of Imagination

Generally speaking the imagination – *Einbildungskraft*- for Kant is a capacity or *power* of the human mind that plays a central role in human experience and is a necessary condition of it. In fact, it is for Kant the very fundament of human cognition. It is also the faculty that gives our consciousness of cognition form and feeling. In what follows I will examine these aspects of the imagination, and their relation to each other.

At the end of the introduction to the second edition of the *Critique of Pure Reason* Kant states that "...there are two stems of human cognition, which may perhaps arise from a common but to us unknown root, namely sensibility and understanding" (B29 – 30). In the "Deduction of Pure Concepts of Understanding", in the first edition of the *Critique of Pure Reason*, Kant calls imagination along with sense and apperception, one of the three "subjective sources of cognition that make human experience in general possible"(A115). As a power of the mind it is distinguished from other powers and capacities in terms of the function it serves in the construction of our overall experience. In one passage in the heart of the revised B Deduction of the first *Critique*, Kant defines the imagination quite simply as follows:

> *Imagination* is the faculty for representing an object even *without its presence* in intuition. (B151)

In his important treatment of the role of imagination in the first *Critique*, Michael Young makes clear that although this passage taken by itself suggests that for Kant, imagination is "simply the capacity for the mental imaging of absent objects" attributing this view to Kant is a gross misunderstanding of his theory of imagination. On Young's account,

> [Kant's] view is rather that imagining involves two moments: immediate sensory awareness, or empirical intuition, and the taking or construing of that awareness as the awareness of something other, or something more, than what immediately appears.[1]

---

1 J. Michael Young, "Kant's View of Imagination" *Kant Studien*, 1988, Volume 79, 140 – 164, pp. 140, 142.

What follows is in substantial agreement with Young's characterization, which correctly relies heavily on Kant's view that "the imagination is a necessary ingredient of perception itself,"[2] Young's analysis also makes clear that what Kant calls *"Einbildungskraft,"* (literally, the power of imagination) is two-pronged, belonging to both stems of human cognition. Finally, Young argues that imagination involves an activity that he calls "construal" or "taking as" that functions actively "in accordance with a rule" to represent a given content *as* some particular thing, without conflating this activity with the representation of the function of the understanding that provides the rule.[3] All of these aspects of Kant's notion of imagination need to be highlighted in making sense of the centrality and the opacity of this notion within Kant's system.

In the first section of the paper I will consider the question of the relative importance of the imaginative synthesis to Kant's overall characterization of the necessary conditions of possible experience, especially in light of his tendency to portray the imagination shrouded in a certain amount of mystery, and in this context I will also examine the claim, made most prominently by Paul Guyer, that Kant relegated the theory of time-determination (and with it of the imaginative synthesis a priori) to a subsidiary part of the argument in the "Analytic." I go on to argue in section II that this view does not take into account the specific role of the imagination in the regulation of the affinity of the manifold. In section III I will argue that Kant's simple definition need not be discarded, but that in an important sense *all* uses of the imagination involve representation of some object, *broadly construed*, even without its presence. Indeed, in the reflective use of imagination this representation of certain ideal objects occurs precisely because they *cannot* be presented.

I

Kant's simple definition of imagination is not to be taken out of context. At B151–52 he elaborates upon it immediately by assimilating the faculty of imagination to both sensibility *and* the understanding: The imaginative faculty is sensible ("belongs to sensibility") he says, because it is can only operate within the medium of intuition, which for human beings is always *sensory* intuition, and al-

---

2 *Critique of Pure Reason*, A120n.
3 J. Michael Young, "Kant's View of Imagination" 1988, Volume 79, p. 153. He also distinguishes this activity from judgment, i.e., from representing something *as* following from that rule, which activity requires according to Young, "reflective criticism" and thus belongs to understanding.

ways only given in the form of space and time.⁴ At the same time he insists that although imagination works under merely receptive spatiotemporal sensible conditions, its *activity* can take place prior to any sensible experience, as an operation *of the understanding a priori* that first makes sense experience as a whole possible by bringing it to the unity of a consciousness:

> insofar as its synthesis is still an exercise of spontaneity, which is determining and not, like sense, merely determinable, and can thus determine the form of sense *a priori* in accordance with the unity of apperception, the imagination is to this extent a faculty for determining the sensibility a priori, and its synthesis of intuitions, *in accordance with the categories,* must be the transcendental synthesis of *imagination,* which is an effect of the understanding on sensibility and its first application (and at the same time the ground of all the others) to objects of the intuition that is possible for us. (B151–52)

Much is contained in this dense passage, but one thing is immediately and remarkably clear, namely that for Kant, imagination has two aspects: on the one hand it is *acting*, and (not merely) *acted upon, determining* and (not merely) *determinable,* spontaneous and creative (productive) and not merely mechanical (reproductive). It is a spontaneous function of mind that conditions possible experience and not *merely* a reflex of body. Yet on the other hand, it also "belongs" to sensibility "on account of the subjective condition under which alone it can give a corresponding intuition to the concepts of understanding." That is, due to the fact that human intuition issensible, and receptive only spatially and temporally, imagination works under the condition of space and time, and its products must be figural. Kant calls this synthesis: "figurative" (synthesis speciosa)" (B151). The imagination specifies sensibility *a priori* in pure spatial and temporal form. This distinguishes the imagination's synthesizing activity from the activity of the "mere" understanding that, according to Kant, relates the pure concepts of the understanding to objects of intuition in general in a "merely purely intellectual" synthesis that involves no consideration of spatiotemporal conditions. The difference is that the intellectual synthesis merely conceives the idea of something in general, while the productive imaginative synthesis specifies or "figures" the unified, general form possible in space and time. This is why the imaginative synthesis is an "effect of the understanding on sensibility and its first application"(B152).

---

4 With regard to "intuition in general" Kant suggests that other kinds of intelligent beings might have different kinds of intuition, either nonsensible (intellecual) intuition, or "some other but still sensible one" (B150), thus allowing that sensibility that is other than spatiotemporal is possible. It would of course be inaccessible to us, and hence irrelevant to critical inquiry into human knowledge.

Kant's "dual aspect" account of imagination suggests interaction between the two "stems" of human cognition. That is, imaginative activity *a priori* is fundamentally constrained by sensibility and at the same time affects sensibility, creating perception by bringing sense data to consciousness[5] (A120n). The fact that these two completely distinct "sources" of human knowledge interact is a fascinating aspect of Kant's theory, and yet one that is hard to assess. For the reader interested in getting to the core of Kant's account of *Einbildungskraft,* it is frustrating that in crucial passages where Kant speaks of the fundamental importance of this faculty he resorts to hand-waving descriptions or makes comments that seem to discourage further attempts at clarifying its source. A prime example of this is at A78/B104, where he provides an account of synthesis and introduces the *Table of Categories,* along the way defining imagination as an unconscious ("blind") process:

> Synthesis is the first thing to which we have to attend if we wish to judge about the first origin of our cognition. Synthesis in general is, as we shall see, the mere effect of the *imagination, of a blind though indispensable function of the soul without which we would have no cognition at all, but of which we are seldom even conscious.* (CPR A78/B103, emphasis added)

As we saw, in the second edition of the *Critique of Pure Reason,* Kant continued to maintain that imagination is the "first origin" of cognition. Here Kant describes imagination's functioning as "blind" and its product as a "mere effect." Presumably the reason Kant calls synthesis in general the *mere effect* (*die bloße Wirkung*) of imaginative activity is that the latter occurs spontaneously, and mostly unconsciously, without reflection or intention on our part. Similarly, in the A Deduction, where Kant defines the understanding as the "the unity of apperception in relation to the synthesis of the imagination" and the *pure* understanding as "this very same unity, in relation to the *transcendental synthesis* of the imagination" (A119), he is vaguely unsettled by the fact that all cognition thus depends upon the transcendental use of this "blind" function:

> It is indeed strange (*befremdlich*), yet from what has been said thus far obvious, that it is only by means of this transcendental function of the imagination that even the affinity of appearances, and with it the association and through the latter finally reproduction in accordance with laws, and consequently experience itself, become possible, for without them no concepts of objects at all would converge into experience. (A123)

---

5 CPR A120n: Here Kant states that his view differs radically from earlier views by taking imagination to be a part of perception: "No psychologist has yet thought that the imagination is a necessary ingredient of perception itself."

At yet another crucial passage in which the imaginative power functioning *a priori* takes center stage we find Kant again treating the imagination's functioning as something that, if not completely mysterious is nevertheless intrinsically difficult to characterize. Discussing the imaginative synthesis that produces empirical schemata for mapping "pure shapes in space" to actual given intuitions he calls it

> a hidden art in the depths of the human soul whose true operations we can divine from nature and lay unveiled before our eyes only with difficulty. (CPR A141/B180–81)

This passage occurs in the schematism chapter where Kant raises the question of how pure concepts of the understanding, which are by their origin completely distinct from intuition can nevertheless be found to apply to sensible appearances[6] Kant's answer is that an intuitive analog to the logical form of pure concepts, viz. figurative forms of time determination *a priori,* are produced *a priori* by the imagination. This imaginative procedure manifests as a set of unique ways of patterning (or "figuring") temporal relations to match or correlate with the unique logical structures of each pure concept. Each category of the understanding is fitted for experience with its correlative time-ordering, or "schema" and Kant attributes this tailoring of an interface between sensibility and understanding to the work of the transcendental imagination:

> The schema of a pure concept of the understanding [...] is[...] only the pure synthesis, in accord with a rule of unity according to concepts in general, which the category expresses, and is a transcendental product of the imagination, which concerns the determination of the inner sense in all representations, insofar as these are to be connected together a priori in one concept in accord with the unity of apperception. (CPR A142/B181)

Kant then launches into a brief summary presentation of the products of the transcendental imagination category by category, saying that he will do this in lieu of "a dry and boring analysis of what is required for transcendental schemata of pure concepts of the understanding in general" (A142/B181). However scintillating the nature of the text that follows, it is disappointing for those interested in hearing more about the imagination's *general* process of schema production *a priori.* Yet given his doubts about the transparency of that process to critical in-

---

6 Whether Kant needs to answer this question here has been the subject of much debate, since in the B-edition version of the Transcendental Deduction he suggests he has resolved the issue by showing that the categories themselves are conditions of the very possibility of any experience whatsoever. See Guyer (1987), *Kant and the Claims of Knowledge*, Cambridge: Cambridge University Press, pp.157 ff.

quiry, it is not surprising. It is hard not to conclude that Kant skirts the issue of what is required in general for the transcendental schemata not because it is dry and boring, but because the critical enterprise is simply not capable of exhibiting that process, let along analyzing it. What *is* open to critical analysis lies on the side of conceptual form, not on the side of unconscious activity that Kant views as "blind" and "deeply hidden" in the human soul.

Paul Guyer dismisses Kant's "hidden art" trope as "creating a sense of obscurity" and argues that "such pessimism seems unwarranted by the official program for the [schematism] chapter" which is to reintroduce the theory of time-determinations produced in imagination as correlates of the pure concepts. He argues that in the Deductions argumentative problems of Kant's own making arose when Kant set aside his theory of the a priori time determination of the pure concepts of understanding as the ground of experience in favor of arguments from the necessity of the synthetic unity of apperception for the possibility of any cognition whatsoever. In Guyer's view "the deepest mystery about the schematism is just the general mystery of why Kant ever departed from his original conception of the direct deduction of the categories from the conditions of time-determination..." in the first place, thus forcing him to return to those conditions, and the role of the transcendental synthesis of imagination in producing them, immediately after the Deductions.[7] Guyer solves the "mystery" based on his earlier analysis of the transcendental deduction:

> Kant's invention of the deduction directly from the apriority of judgment, and even more that of the deduction from the a priori certainty of apperception, led him to reconceive of the theory of time-determination as a subsidiary part of his theory of knowledge, separate from the transcendental deduction itself. (Guyer, 160)

Guyer is right to point out that Kant's introduction of an argument from the unity of apperception in the Deductions deflects attention away from the theory of time-determination, which establishes the objective validity of the categories based on the synthesis of pure temporal orderings that pattern their logical structure. This account, laid out in the schematism and the Principles chapters, and especially in the arguments of the second analogy, is by far the more persuasive approach for many of his readers. Furthermore, it has long been alleged that Kant demotes the role of imagination in the B-edition Deduction to a subordinate function of the understanding, and Guyer's claim that Kant treats time-determination in the argument of the deduction as "a merely subsidiary part of his theory of knowledge" (160) would appear to bolster this allegation. Given that time-

---

7 Guyer, Paul (1987), *Claims of Knowledge*, p. 158 & p. 160.

determination is the defining function and ultimate *product* of the transcendental imaginative synthesis, it is easy to conclude that Kant does indeed recoil from the imagination as a fundamental faculty.[8] Especially in light of Kant's evident hesitation to say much at all about the actual process of imaginative functioning, it is easy to conclude that Kant backs away from placing much argumentative weight on arguments for the validity of the categories that rest too heavily on the transcendental synthesis of imagination.

Nevertheless even with the introduction of the principle of apperceptive unity, nothing could be further from the truth. In the following section I will argue that the transcendental synthesis in imagination remains at the heart of his account of cognition in both Deductions, not subordinated, but encased, as it were, within his account of the transcendental unity of apperception as "the highest principle of all use of the understanding."

2

Kant's comments about the obscurity of the transcendental synthesis notwithstanding, he actually has quite a bit to say about it in the Deductions under the heading of "figurative" synthesis. In its most fundamental, *a priori* use, imaginative synthesis is *the* ursynthetic mental activity that *determines* experience *a priori*, by an activity that involves the production of pure temporal and spatial forms, i. e., basic "figures" in space and time. Time-orderings in particular are the focus of the *a priori* imaginative synthesis because time is the form of inner sense and the medium of imagination. But if it is clear why the imagination works with temporal forms, it is not as clear why it must be constrained to look to the pure concepts for forms of judgment, as opposed to, say, simple associative principles, in giving temporal form to experience. The role of the principle of the synthetic unity of consciousness, for Kant seems to be that of an overarching principle of the understanding that jumpstarts the imaginative process by constraining it to look for more than merely subjective (associated) time orderings in inner sense.

---

8 Heidegger's criticism of this is the *locus classicus*, in *Kant and the Problem of Metaphysics*, Bloomington: Indiana University Press, 1997. See paragraph 17 at B136: "The supreme principle of the possibility of all intuition in relation to sensibility [is] that all the manifold of sensibility stands under the formal conditions of space and time. The supreme principle of all intuition in relation to the understanding is that all the manifold of intuition stand under conditions of the original synthetic unity of apperception.

This overarching principle, Kant argues, is that of the unity of apperception: the requirement that all cognition take place in a single, unified consciousness. As mentioned, Guyer criticizes Kant's argument in the Deduction for relying so heavily on the assumption of the strong claim that all perception presupposes *apperception*. But in Kant's discussion of the relation of the transcendental unity of apperception to the transcendental synthesis of imagination in the third section of the first edition deduction (A115), Kant lists apperception along with imagination and sensibility as just one among equals: "the three subjective sources of cognition." Analyzing the conditions of the possibility of experience "from above," Kant expresses the synthetic unity of apperception as a transcendental *principle* that states that:

> We are conscious *a priori* of the thoroughgoing identity of ourselves with regard to all representations that can ever belong to our cognition, as a necessary condition of the possibility of all representations. (CPR A116)

Kant means by this *not* that we are empirically conscious of this identity at all times, but that we are always *capable* of such consciousness. He calls this capacity, variously, "transcendental consciousness" or "pure apperception", and the claim that "We are conscious *a priori* of the thoroughgoing identity of ourselves with regard to all representations that can ever belong to our cognition, as a necessary condition of the possibility of all representations" Kant calls "the transcendental principle of the synthetic unity of the manifold in all possible intuition."[9] He then immediately claims that this synthetic unity "presupposes or includes" the productive, *a priori* synthesizing activity of the imagination, and concludes that the principle of this synthesis is "prior to apperception" and therefore "the ground of the possibility of all cognition, especially that of experience." (A118).

Finally, as we saw earlier, he concludes that the understanding just *is* the power of imagination plus its product, apperception:

> the unity of apperception in relation to the synthesis of the imagination, and this very same unity, in relation to the transcendental synthesis of the imagination, is the pure understanding. (A119)

---

9 A117. See also A116n: "But it should not go unnoticed that the mere representation *I* in relation to all others (the collective unity of which it makes possible) is the transcendental consciousness. Now it does not matter here whether this representation be clear (empirical consciousness) or obscure, even whether it be actual; but the possibility of the logical form of all cognition necessarily rests on the relationship to this apperception *as a faculty*.

Hence the A Deduction already establishes an intimate connection between imagination, apperception and understanding that asserts, as it were, the genetic priority of the power of imagination as the activity that begets all experience. Kant then defines the categories of the understanding as "pure *a priori* cognitions that each contain the necessary unity of the pure synthesis of the imagination in regard to all possible experience" (A119), which means that for Kant, the categories are the rules or functions that provide the power of imagination with formal rules of synthesis for a unified consciousness. This means that the imagination's productive synthesis *a priori* – the very synthesis that is at the root of apperception – is nothing more nor less than the process of fitting categorical structures with intuition – it is schematizing, providing time-determinations *a priori* for each and every category. In this respect, time-determination is already assumed (but not yet explained) in Kant's Deduction, and along with the imagination which enacts it, is anything but subsidiary in the Deduction.

Having arrived at the transcendental synthesis of imagination by starting "from above" with the unity of apperception, he returns to examine the process "from below" beginning with the work of the empirical imagination *a posteriori*. He describes imagination as empirically active in the apprehension of the manifold of sensory intuition before it can bring it into an image, and therefore as productive or, as he puts it, as "a necessary ingredient of perception itself."(A120n) It is further active, "reproductively" in exhibiting whole series of perceptions from which to recall past ones, and further associating them in various combinations. But, and this is the crucial point against Hume, mere association would not lead to cognition if there were not some guarantee that what the imagination attempts to associate is indeed *associable*. As Kant puts it, without the *associability* of perceptions,

> a multitude of perceptions and even an entire sensibility would be possible in which much empirical consciousness would be encountered in my mind, but separated, and without belonging to *one* consciousness of myself[.]. (A122)

He then goes on to argue that the *possibility* of association depends upon an "objective ground", i.e, *universally necessary law* that *regulates* and guides the imaginative process a priori.

> There must therefore be an objective ground, i.e., one that can be understood *a priori,* antecedently to all empirical laws of the imagination, on which rests the possibility, indeed even the necessity of a law extending through all appearances, *a law, namely for regarding them* throughout as data of sense that are associable in themselves and subject to universal laws of a thoroughgoing connection in reproduction". (A122)

He labels this a priori associability the "affinity" of appearances, and the law that describes it is what in the B Deduction he calls the "highest principle of all use of the understanding" i.e., the principle of the unity of apperception. That is, the principle of the synthetic unity of apperception is the principle that articulates this affinity, and it is the transcendental imagination whose function it is to produce it *a priori*.

In the context of Kant's later work on reflective aesthetic judgment in the third *Critique* it is striking to see Kant, in the first *Critique*, and at the heart of the Deduction, invoking what appears to be a *regulative* principle that directs the imagination to "regard [appearances] throughout as data of sense that are associable in themselves and subject to universal laws of a thoroughgoing connection in reproduction." Taking Kant at his word here clearly changes the status of the principle of apperception to a heuristic, or reflective principle *a priori*, necessary for cognition, but not constitutive of it in the same way that the schematized categories and the principles of pure understanding are. Rather it is a fundamental rule for the direction of the imagination to seek a framework for order and connection by seeking to fit what is given in intuition to what the mind already possesses by way of cognitive structure in the categories. It is then in the very process of seeking, finding and producing this fit that the imagination actually *creates* associability by producing time-determinations in accordance with categories.

Fleshing out the details of the imaginative process of time-determination is left for the schematism and the Principles chapters, but the process itself is very much presupposed by and included in Kant's account of the unity of apperception in the Deduction. Thus one response to Guyer's question of about why Kant "departed from his original conception of the direct deduction of the categories from the conditions of time-determination" is that he in fact did *not* depart from his original argument at all. To the contrary, far from departing from the argument from time-determination, and further still from subordinating it to "a merely subsidiary part of his theory of knowledge," Kant argues, in both deductions, for the *priority* of the transcendental synthesis of imagination to the transcendental unity of apperception, claiming that the imaginative synthesis is to be found already "contained in" the categories.[10] This suggests that it is the categories that are rendered "associable" (schematized) by the transcendental synthesis of imagination, and so are themselves "presupposed or included" in the unity of apperception as the "objective ground of all association of appearances" that he calls "their *affinity*" (A122). It is certainly compatible with this account that

---

10 Guyer, Paul (1987), *Claims of Knowledge*, p. 160.

they are the fundamental concepts that give representations their affinity in the overall cognition of the subject.

Thus, a closer look at the role of the transcendental synthesis of the imagination in Kant's deductions makes clear that he did not relegate the theory of time determination to a mere subsidiary to the theory of the unity of apperception in the Deductions. What he did do, I suggest, was to switch his focus in the Deduction to the higher order, *regulative* contribution that the understanding as a purely discursive faculty of rules makes to the transcendental synthesis of imagination that lies at the heart of all cognition and constitutes the interface of sensibility and understanding. That contribution, based on the need for a unified consciousness as the backdrop for all cognition, is merely a regulative principle that guides imagination to "regard appearances throughout as data of sense that are associable"[11]

In the end, however, it is also true that Kant's reference to the schematism as a "hidden art" in the depths of the human soul is an expression, as Guyer puts it, of a certain pessimism on Kant's part about explaining the mechanism, or "art" of a faculty that straddles the divide in human nature between sense and concepts, intuition and understanding, and receptivity and spontaneity. In detailing, in the "schematism" and "Principles" the role of time-determination of pure concepts of the understanding by the imagination *a priori*, Kant reaches the limits of critical analysis. The process and products of imagination do an enormous amount of work in his account of cognition in general, but imagination presents him in the end with the methodological difficulty, if not the methodological impossibility, of characterizing its most basic *processes* via critical inquiry. Imagination's products can only be described in critical terms via an account of rules of the understanding that must be in place for the direction of imaginative activity *a priori*. A regulative rule for finding the unity of consciousness and constitutive rules for determining the manifold of intuition within consciousness can be elucidated by transcendental procedures and Kant does so, with varying degrees of success, in the Deduction, Schematism and Principles chapters of the first *Critique*.

But a complete account of the actual *functioning of the imagination* in the production the affinity of the manifold and of time-determination a priori would need to clarify the point of contact and the interaction of the sensible

---

11 This also explains the sense in which the concept associated with the principle of the unity of apperception is for Kant a purely formal, empty concept of an "I". For an excellent account of the sense of self identity that Kant is working with in the deductions, and that also construes the unity of apperception as regulative, see José Luis Bermúdez , "The Unity of Apperception in the Critique of Pure Reason", *European Journal of Philosophy*, Vol. 2: 3, Dec. 1994, pp. 213–240.

and intellectual aspects of imaginative synthesis a priori, and this is a task that cannot be performed by transcendental deduction. As Kant sees it this is a question of fact – a "physiological derivation"(A86 – 87/B119).[12] It is a question about the very constitution of human nature that must remain opaque to critical reflection. As Kant himself admits early in the Transcendental Analytic, his task is to analyze

> the faculty of understanding itself, in order to research the possibility of a priori concepts by seeking them only in the understanding as their birthplace and analyzing its pure use in general, for this is the proper business of transcendental philosophy.(CPR A65 – 66/B90)

The problem, for Kant is that imagination merely a child of the understanding, that "gestates in sensibility", and this hybrid faculty, operating according to rules of the understanding but directly affected by matter from the senses (A86-B118) sits uneasily within the overall critical enterprise. Critically speaking it is not imagination, but reason itself that is blind, or at best severely myopic, when it comes to this aspect of our experience. Imagination is "hidden" from critical view at the very origins of the process that first sets the cognitive apparatus into motion.

3

> It is indeed strange (befremdlich), yet from what has been said thus far obvious, that it is only by means of this transcendental function of the imagination that even the affinity of appearances, and with it the association and through the latter finally reproduction in accordance with laws, and consequently experience itself, become possible, for without them no concepts of objects at all would converge into experience. (Critique of Pure Reason, A123)

Given the centrality of the imagination as the root of all cognition in the *Critique of Pure Reason*, the lack of reference to the imagination in the second *Critique* is striking – all the more so since here he begins in earnest to refer to moral judg-

---

12 See also the *Anthropology from a Pragmatic Point of View*, Immanuel Kant: Anthropology, History, and Education, ed. Günter Zöller and Robert B. Louden, Cambridge: Cambridge University Press, 2007, pp. 231 ff. (VII: 119 ff.). Kant's authorized collection of lecture notes for more on what he labels "physiological" methodology.

ment as a kind of cognition [*Erkenntnis*]. Its absence can of course be explained in light of Kant's reduction of moral value to the goodness of the will: A moral judgment for Kant expresses a decision to act in a certain way, viz., in accordance with the categorical imperative, and as is well known, for Kant the morality of an action consists in the determination of the (good) will to perform it, *regardless of whether the objective of the act can in fact be realized by the agent.* In other words, a moral judgment, for Kant, is moral insofar as it involves a consciously determined choice *necessarily* to act in accordance with the moral law regardless of the actual outcome of the act. Assuming, as Kant does, that the moral law is both rational, a priori, and experienced by us as a basic "datum" of our humanity, it makes sense for him to say that the purely moral judgment does not in itself involve calling to mind – i.e., imagining – various possible outcomes of the action it determines. In a fundamental sense moral judgment must be independent of contingent circumstances, while imagination, in an equally fundamental sense, must always work with contingencies: even when it synthesizes *a priori*, the imagination synthesizes that which is contingently given to it[13] Kant recognizes, however, that categorically action guiding, *purely* moral judgment is not the only kind of practical judgment. That is, it is not the only kind of judging that determines an action. Kant spells this out in some detail in the *First Introduction to the Critique of Judgment*, where he claims that there are two kinds of practical propositions, only one of which is irreducibly, or "purely" practical, and hence alone belongs to the "special part of a system of rational cognitions" that he calls "practical philosophy."

Irreducibly practical propositions are pure: they present [*darstellen*] the necessity of an act based purely on the form of the law that determines it, without regard for the means of producing the object the act aims to produce. That is, for Kant a proposition is irreducibly practical if it expresses an "ought" categorically, thereby commanding that we do something without regard to whether or not the outcome is guaranteed or even possible in nature. They are in his words,

> practical propositions that directly present as necessary the determination of an act by the mere representation of the act's form (in terms of laws as such) without regard to the means used to achieve the act's object. (XX:199 FI)

---

13 In the final analysis there is contingency in the very fact that we have, a priori, the kinds of intuition and categories of experience that we do: B145–46: "But for the peculiarity of the understanding, that it is able to bring about the unity of apperception *a priori* only by means of the categories and only through precisely this kind and number of them, a further ground may be offered for why we have precisely these and no other functions for judgment or for why space and time are the sole forms of our possible intuition."

The object of the action determined by a pure practical proposition (that is, by a purely moral judgment based on the idea or "principle" of our freedom) is not directly a result of a desire for that object, but rather a consequence of the moral principle itself. Moreover, the moral law only enjoins the achievement of one object: that of the highest good possible in the world. That object is a mere ideal to be hoped and striven for; it is not a natural object, and hence cannot be *known* to be possible. Hence the "ought" of the practical imperative that one ought to strive to bring about the highest good within the realm of nature is a pure practical command. Indeed, Kant says, it is the *Categorical Imperative.*[14] This leads him to ask, in the second *Critique*, how it can be rational to command of ourselves an action whose realization – or even approximate realization – might well be impossible. This is a major problem for Kant's *Critique of Practical Reason* that he famously attempts to solve with his invocation of the Postulates of practical reason asserting the existence of God and immortality. No knowledge of their truth is possible, he says, but nevertheless, in practice, postulating these two pillars of Christian monotheism blocks the threat of rational moral despair.

Yet even if Kant had been fully convinced that he had solved the problem of how reason can legitimately command us to hope and strive for the highest good, he remained uneasy about the gap in human experience that this rigid separation of moral practice from theoretical knowledge created.[15] For this reason in the third *Critique* Kant introduces another kind of practical judgment that is reducible to a theoretical claim about what is actual and/or possible in nature, and he allows that although these are distinct from purely practical moral judgments, they cannot be entirely irrelevant to them. Clearly, morality and moral judgment require the ability to recognize relevant and salient aspects of a moral context.[16] Such recognition requires reflection upon and comparison of various cases, both of which are cognitive activities that employ the imaginative capacity for representing what is not present. Kant therefore cannot deny that within the overall framework of moral judging, many practical judgments

---

14 Critique of Practical Reason, trans. and ed. by Mary Gregor, Cambridge: Cambridge University Press, 1997.

15 See Critique of the Power of Judgment, trans. Paul Guyer and Eric Matthews, Cambridge: Cambridge University Press, 2000, second Introduction, V:175–76. Kant also manifests concern for this gap in the Methodology section of the second Critique in his account of the aesthetic exercises to be used to lead children to mature moral judgment.

16 See Barbara Herman: *The Practice of Moral Judgment*, Cambridge, Mass.: Harvard University Press, 1993.

are involved that are not themselves *purely* practical.[17] In the *First Introduction to the Critique of Judgment*, Kant describes such judgments as involving a determination to act in certain ways, for example, so as to test a hypothesis (i. e., to produce an experiment) or *so as to invoke in ourselves a certain state of mind*. The difference, he says, is that

> the principles we follow in performing experiments [in physics] must themselves always be obtained from our knowledge of nature, and hence from theory. *The same holds for the practical precepts concerning the voluntary production in us of a certain state of mind (e. g., the state of stirring or restraining our imagination, pacifying or abating our inclinations...)*...(XX:198 – 99, emphasis added)

The last claim introduces an interesting kind of practical judgment to Kant's overall account of that faculty, namely "practical propositions that assert the possibility of an object through our power of choice" (XX: 199). Such judgments in which we more or less consciously act in certain ways in order to control our own state of mind follow principles "obtained from our knowledge of nature." As such they are not rooted in the principle of our own freedom and are not therefore "purely" practical. Rather, they are only instrumental or in Kant's terms, "technical" judgments. That is, they are "how to" precepts derived from our knowledge of the objects they aim to bring about. In addition to prior scientific knowledge, such judgments might rely, for instance, on prior experimentation, or in the case of *"practical precepts concerning the voluntary production in us of a certain state of mind,"* they might rely also on first personal experience of what kinds of mental *activities* produce a certain mental *state* in ourselves.

The fact that these judgments are technical rather than purely practical does not entail, however, that "technically practical" judgments play no role in setting the stage for and maintaining the conditions of purely practical moral judgment. Indeed, if certain kinds of technical judgments contribute to conditions of the human mental environment without which moral judgment could not be developed and sustained, the cognitive faculties, including the imagination that enacts these judgments must be seen as necessary to the very existence of human morality. Even if these practices of judgment cannot be specified fully *a priori* because they are fundamentally constrained by the empirical facts of the world, they could still produce in the world the very conditions under which alone *pure* practical judgment may be realized. In this sense, to para-

---

17 See Bernard Freydburg's spirited defense of the pervasive role of imagination in the second *Critique: Imagination in Kant's Critique of Practical Reason*, Bloomington: Indiana University Press, 2005. See also Robert Louden's *Kant's Impure Ethics*, Oxford University Press, 2000.

phrase Kant's comment on imagination, it appears strange, but obvious, that pure practical judgment can take root, develop and sustain itself only by means of this *technical* function of judgment.[18]

The two examples of technical judgment that we see mentioned by Kant are not randomly chosen. In the context of the *Critique of Judgment* Kant wants to give an account of the a priori principle that underlies two important aspects of experience not yet dealt with in the critical theory: the uniformity and systematicity in nature's laws and forms that scientific inquiry depends upon, and the harmony of natural purposes with the ends of human moral agency. The latter is a crucial part of Kant's systematic aesthetics, and is central to the question of imagination's role in human moral judgment.

Thus Kant's introduction of technical judgments about how to produce in ourselves a certain state of mind is extremely important: We learn from direct observation and the study of nature that some aesthetic states of mind – for example, feelings of pleasurable disengagement from everyday concerns during the free play of imagination, or feelings of quietude and serenity when desires are calmed– are generally more conducive to moral determination than others. Technical judgments that involve *"a certain state of mind ... the state of stirring or restraining our imagination, pacifying or abating our inclinations"* are the judgments required for maintaining our purely practically motivated desire to do the right thing, and to make the world a morally better place.

Given the fundamental role that imagination plays in all cognition for Kant, the use of technical rules guiding mental activity for the purpose of producing or maintaining a mental state will surely depend upon the imagination "stirring or restraining" itself in accordance with these technical precepts. In the case of moral judgment then, the question is *what* rules of judgment, and *what* corresponding activity of the imagination, will be invoked as the scaffolding, as it were, for the growth and maintenance of a sound moral system that is sustainable in the actual world? In the remainder of this paper I will briefly sketch how Kant adapts his first *Critique* account of the imagination's unconscious production of a transcendental affinity or associability in intuition to explain the *conscious, reflective* activity of imagination that is directed at changing our state of mind in judgments of taste. I will conclude with a brief discussion of Kant's

---

18 For this reason, theories that heavily emphasize the "primacy of the practical" in Kant's architechtonic overlook a crucial aspect, namely the degree to which Kant must assume an interplay of the cognitive and the practical in his theory of the power of judgment. See Jane Kneller, *Kant and the Power of Imagination*, Chapter 4 (Cambridge University Press: 2007), for an extended discussion of this point in light of current scholarship asserting the primacy of practical reasoning in Kant's philosophy.

views on how imaginative activity, guided by a technical principle aimed at exhibiting the possibility of harmonized purposiveness between the nature and human morality, develops and maintains moral judgment.

4

> For in the power of judgment,
> understanding and imagination are considered
> as they relate to each other. That relation can be
> considered objectively (as was done in the
> transcendental schematism of the power of
> judgment) as belonging to cognition; but one can
> also consider this same [eben dieses] relation
> between the two cognitive powers merely
> subjectively, as one helps or hinders the other in
> one and the same representation and thereby
> affects the state of mind, and [is] therefore a
> relation which is **capable of being sensed**
> [empfindbar] (which is not the case in the
> separate use of any other faculty of cognition).
> (XX:223, Kant's emphasis)

In order to explain the indirect but necessary role of the power of imagination in moral judgment, it is important to take Kant at his word when, as in the excerpt quoted above, he asserts that the relationship of imagination and understanding in reflective judgment introduced in the third *Critique* is *the same as* the relation of imagination and understanding in the first *Critique* account of the unity of objective experience made possible for us by the transcendental synthesis of imagination in the schematism of the categories. In the cognitive case, as we saw earlier, the imaginative power, following a regulative demand that appearances *in general* be regarded as associable (the principle of the unity of apperception), seeks *a priori* an affinity in the manifold appearances. The power of imagination thus first makes a world of objects possible. It does not thereby create that world, of course. Kant is quite clear that associability in the end is still contingent on the fact that the "given" in appearances is not so disparate and confused that the imaginative function would simply fail to find connections, in which case no affinity for intuitive sensory beings like ourselves would be constituted. Nevertheless the affinity that the imagination does succeed in finding then *constitutes* the objective nature of human individual cognition by, as we saw, schematizing it in a causal nexus according to time determinations (the categories) *a priori*. How can this unconscious yet cognitive activity of the imagination in concert with laws of the understanding, be understood as "the very same [*eben dieses*] relation", as the self-consciously reflective activity of the imagination indeterminately at play with representations of the understanding, or as Kant puts it, with "the two cognitive powers [considered] merely subjectively"?[19]

In the third *Critique* Kant argues for a regulative, *a priori* principle of judgment that directs imagination to seek a systematic connection of particular laws of nature and he also argues that this same regulative principle (the principle of the formal purposiveness of nature) when prompted by the aesthetic formal properties of objects, guides the a free activity of the imagination in concert with the understanding. This activity produces in us mental state that he describes as a *feeling* of purposiveness without claim to knowledge of purposiveness. Yet because this feeling is based on cognitive powers that everyone shares, the feeling itself has what Kant calls "subjective universality" and is, given the proper circumstances, capable of being universally shared.[20] This is the basis for the well-known "deduction" of judgments of taste, ie., judgments of beauty. It is also part of Kant's larger project in the "Critique of Aesthetic Judgment" of arguing for aesthetic reflective judgment as a means to the development of moral judgment.[21]

---

19 Kant makes the same assertion for the unity of empirical laws in the sciences, and the same question of how this ursynthetic activity can understood to be "one and the same" as the unity of "the affinity of particular laws under more general ones, [that] qualifies for an experience, as an empirical system" can be raised here: *For **unity of nature in time and space**, and unity of the experience possible for us are one and the same [einerlei]...thus it is a subjectively necessary **presupposition** that...nature itself, through the affinity of particular laws under more general ones, qualifies for an experience, as an empirical system. (XX:209, Kant's emphasis)*. Seeking systematicity of nature in its laws and forms makes possible the discovery of those laws and forms possible – if they are there to be discovered. *First Introduction to the Critique of Judgment* in *Critique of Judgment*, trans. Werner S. Pluhar (Indianapolis: Hackett Publishing Co., 1987).
20 Ibid.
21 See for instance V:354, where Kant argues that judgments of beauty are a kind of bridge to morality, and section 42, (V:300–301) where he claims that aesthetic reflective judgments of beauty have an affinity with moral feeling and give rise in us to a moral interest. Moreover, he argues that the very activity of imagination acting in free, non-determining harmony with objects of the understanding produces feelings of "disinterested" pleasure that are themselves of great moral interest to us: for the sense of harmony of our natural forms with our cognitive faculties in general – the feeling of their very suitability for stirring our imagination to pure reflection – constitute a sign of the possibility of a world of nature that is in harmony with our specific subjectivity and to that extent is open to our efforts to change it. Insofar as pure practical reason commands as a duty that we strive to harmonize natural and moral purposes, it can be understood as directing the imagination to seek a systematic connection between the system of rational moral ends of practical reason on the one hand and the natural system within which human beings actually live on the other. If we find the signs of moral purposiveness in

How, is this to be understood as "the very same" relation that exists between the imagination and understanding in cognition? In both cases imagination is guided by a principle for judging (synthesizing or combining the elements of a manifold) for the sake of producing a unity of consciousness. Once constituted through transcendental imaginative synthesis, this unitary consciousness is *able to sense its own activity.* Having a unified consciousness means being able to turn one's mental gaze inward upon oneself and in so doing being able to feel the activity of "that which is merely nature in the subject" acting with and through one's own reason.[22] This feeling of the interaction of the sensory aspect of imagination in concert with the intellectual functions of the understanding is not empirical *knowledge* of the unity of consciousness, but it is a universally communicable instance of *sensing* that unity. It is the feeling of the fit between nature and reason from the inside out, so to speak, and as such constitutes a kind of signal from nature of its harmonious coordination with moral reason.

This harmony is of utmost importance to Kant's ethics, because we are able to reconcile the demands of moral reason upon our natural selves only if the systematic demands of pure practical reason are related in some way to the whole of the natural system within which it is ensconced. Hence imagination, regulated *a priori* by a principle that calls for regarding appearances *in their particularity* as associable, must seek not only uniformity in nature but also an affinity of nature *for our own moral subjectivity.* The combination of the fact that morality requires us to attempt to *realize* the highest good, (that is, a systematic coordination of individual virtues and individual happiness) on the one hand, and on the other, the fact that we have so little evidence of systematic human success in this regard, requires the power of imagination in reflection to produce a moral unity of apperception: a reflective self-consciousness that supports our moral self-confidence and underwrites moral hope.

Kant thus turns to reflective *aesthetic* judgment and the interests to which it gives rise when he addresses the need to reconcile what we must do, morally speaking, with what we may hope for ourselves. The details of Kant's analysis

---

nature that we seek we have reason to believe in our own moral efficacy in the world – and hence to we may be motivated by it. Although this reflective unity is not guaranteed and is contingent upon the circumstances of the individual subject in a way that our cognitive experiences are not, it is also true that the beauty of certain natural forms that we find to be in harmony with our moral commitments produce in us a sense of purposiveness that serves, or could serve, to produce the very efficacy that morality commands us to attempt.

22 *Critique of the Power of Judgment*, trans. Paul Guyer and Eric Matthews (*Cambridge: Cambridge University Press, 2000*), V:344.

of judgments of beauty (judgments of taste) need not be explored here. Rather, taking a cue from Kant's account of practical precepts that concern the voluntary production in us of a certain state of mind, I will end by pointing to the relevance of the fact that for Kant, aesthetic reflective judgments in the "Analytic of Aesthetic Judgment" in the *Critique of Judgment* are technical insofar as they invoke a practical precept governing both the stirring and the restraint of the imagination. Aesthetic reflective judgments produce a certain state of mind, namely a *feeling* of disinterested pleasure from the contemplation of beautiful forms in nature. In such judgments imagination is *limited* by abstraction from interests of the sort associated with immediate, unrestrained appetites, from interests of possession, and even from interests of moral agency. Limited in this way, it is also free to engage in simply "playing" with these forms in imagination. The well-known "free play of the imagination" in Kant's reflective aesthetic judgment is thus guided by a technical precept that tells us how *not* to regard nature, in order to produce a certain state of mind in which we become conscious of our existence as a unified consciousness among other unified consciousnesses. This precept guides and facilitates the continued play of the imagination by reducing the personal interest, including the interest of objective judgment itself, that slows and ultimately stops the pleasurable aesthetic state of mind. Freed from the constraints of "normal" cognition, the imagination is "stirred" and allowed to enter into a purely aesthetic free play. The result is not an objective judgment, but rather a universally communicable judgment that asserts the pleasurable feeling of the harmony of nature and reason within the conditions of cognition itself.

Another way of putting the point is in terms of what imagination as an *a priori* cognitive power can produce. We have seen that it is capable of producing the spatiotemporal forms that give rise to our experience of nature *a priori*, and that guided by the search for unity of consciousness it schematizes or "figures" objects in accordance with the temporal orderings of the categories. The art of bringing about something that we want is a technical skill. Cognitively speaking, as we saw, imagination as an "art" hidden in the depths of the human mind can be said to craft the affinity of the manifold with the guidance of an a priori principle of association: namely the law that manifold appearances must be *regarded* as associable, or as Kant says, regarded as having an affinity with each other such that they can be combined in one consciousness.[23] As a result, coherent individual experience first becomes possible.

---

23 See Samantha Matherne's ms: "for a discussion of the "art" of schematism in empirical judgment".

So too, then, the imagination might be *self-consciously* directed to associate a manifold (of forms) upon the adoption of a principle that regulates the search for associability of these forms. As with the understanding, the law that regulates the faculty of judgment when it functions reflectively – heautonomously – is a rule for how the imagination should regard or construe a given manifold. For the understanding this is the principle of the unity of apperception, which, as we saw, is the "law" or "ground" that "constrains us to regard all appearances as data of the senses that must be associable in themselves and subject to universal rules of a thoroughgoing connection in their reproduction" (A122). Because no experience could be possible at all without such a unification in one consciousness, the resulting "self" is the ground of all perception. schematism, insofar as the imagination brings about the affinity of a manifold via temporal orderings correlated with the categories, is thus an artifact of the imagination that makes perception possible. For Kant, the same activity of seeking, and in so doing producing, this affinity can take place in self-consciousness. Thus it is one and the same power of the imagination that crafts both our cognitive world of experience and that reunites us with this world as moral agents. To the extent that such a reunification is necessary for the possibility of moral motivation, the imagination is indeed strangely but obviously at the root of all human experience.

Emily Brady
Imagination and Freedom in the Kantian Sublime

1 Introduction

Imagination plays an important role in Kant's mature theory of the sublime, yet commentators often characterize this role largely in terms of failure, without recognizing and articulating its distinctive, positive contribution. In this chapter, I discuss how imagination operates in more positive ways and show how its activity is intimately tied up with sublime judgment and feeling.[1] More specifically, I shall argue that in the mathematically sublime, imagination is expanded through attempts to capture the infinite, an activity that can be described in terms of aesthetic freedom, the sublime counterpart, as it were, for the free play of imagination in the beautiful. In the dynamically sublime, we find that imagination functions negatively in being overwhelmed by powerful natural qualities, yet also positively through modes of projection and identification. Imagination's negative and positive functioning is crucial to the feeling of moral freedom which emerges through this second type of sublimity. Through an exploration of the constructive functioning of imagination in the sublime, I aim to extend our understanding of this mental power in Kant's philosophy and to link its activity to different modes of freedom.

Kant's theory of the sublime stands out from those of his predecessors and contemporaries for its strong metaphysical component, which is rooted in an attempt to formulate a conception of the sublime within the context of his tran-

---

[1] Sarah Gibbons also distinguishes the positive role of imagination and the neglect or downplaying of imagination by some commentators (*Kant's Theory of Imagination: Bridging Gaps in Judgment and Experience* (Oxford: Clarendon Press, 1994), 136; 148ff; 150–151 n. 26; chap. 4). She refers specifically to Crowther's views, and suggests that had he recognized the positive role of imagination in the dynamically sublime as well, he might have grasped the dynamically sublime as an aesthetic concept (p. 149). Crowther argues that Kant's views are ambiguous, and uses this as the basis for his attempt to reconstruct what a positive role for imagination might look like. See *The Kantian Sublime: From Morality to Art* (Oxford: Clarendon Press, 1989), 132–134. Rudolf Makkreel's discussion of the regress of imagination in the sublime also supports my argument for a positive rather than only negative role for imagination, see *Imagination and Understanding in Kant* (Chicago: Chicago University Press, 1994), 303–315; chap. 3). Jane Kneller is perhaps more neutral on this point, see *Kant and the Power of Imagination* (Cambridge: Cambridge University Press, 2007).

scendental system and critical philosophy. As such, it links the sublime as a form of aesthetic experience with the sense of freedom possessed by moral beings. However, many of his ideas are indebted to other eighteenth-century discussions of the sublime. In terms of British influences on Kant's aesthetic theory, we know that Kant was familiar with the work of Shaftesbury, Addison, Hutcheson, Hume, Blair, Kames, Gerard, and Beattie.[2] In Germany, Sulzer, Baumgarten, and Mendelssohn have been named as primary influences.[3] Although I will not focus on these theories, my arguments in the paper will draw on some of them in order to better understand imagination's role. This historical context is also important for placing the Kantian sublime within the tradition of *aesthetic* understandings of the sublime. My discussion of imagination and the sublime will be aesthetically-driven, which is to say that, first, I will be concerned with the sublime as an aesthetic concept with a legacy in both British and German aesthetic philosophy; and, second, my focus will be on the *Critique of the Power Judgment*, rather than other writings in Kant's oeuvre. I have argued elsewhere for an aesthetic as opposed to moral interpretation of the sublime in Kant, but the case I build for imagination's role here will also provide some support for this kind of view.[4]

2 See Gracyk, Theodore A. (1986): 'Kant's Shifting Debt to British Aesthetics,' *British Journal of Aesthetics*, 26:3, 204–217.

3 See Guyer, Paul (2007): '18th Century German Aesthetics,' *Stanford Encyclopedia of Philosophy*. http://plato.stanford.edu/entries/aesthetics-18th-german/ (accessed 15/5/12); Guyer, Paul (2005): *Values of Beauty*. Cambridge: Cambridge University Press, 33.

4 See Brady, Emily (2012): 'Reassessing Aesthetic Appreciation of Nature in the Kantian Sublime,' *Journal of Aesthetic Education*, 46:1, 91–109. Those who support an aesthetic reading of the sublime in Kant include, Matthews, Patricia (1997): 'Feeling and Aesthetic Judgment: A Rejoinder to Tom Huhn,' *Journal of Aesthetics and Art Criticism*, 55, 58–60; and (1996): 'Kant's Sublime: A Form of Pure Aesthetic Reflective Judgment,' *Journal of Aesthetics and Art Criticism*, 54; Guyer, Paul (1996): *Kant and the Experience of Freedom*, Cambridge: Cambridge University Press; Makkreel (1994): Others argue that the sublime lies closer to the moral, see: Crawford, Donald (1985): 'The Place of the Sublime in Kant's Aesthetic Theory,' in Richard Kennington, ed. *The Philosophy of Immanuel Kant*, Washington: The Catholic University of America Press; Crowther (1989); Schaper, Eva (1992): 'Taste, Sublimity and Genius,' in Paul Guyer, ed. *The Cambridge Companion to Kant*, Cambridge: Cambridge University Press.

2 Imagination in Judgments of the Beautiful and the Sublime

To introduce the role of imagination in Kant's theory of the sublime, I begin where Kant begins, that is, with an analysis of the similarities and differences between the sublime and the beautiful.[5] Judgments of taste (the beautiful) and judgments of the sublime are both types of aesthetic judgment for Kant, and as such they share a set of features.[6] They are reflective judgments that involve only indeterminate concepts rather than determinate ones and are singular judgments with subjective universal validity. In these respects they are presented as contrasting with judgments of sense and logical, objective judgments, which involve determinate concepts. As such, neither involves a conceptual understanding of the object as a ground, motivation or end. Both types of judgment are disinterested, and characterize an aesthetic appreciation that is distinct from any interest in the uses or function of the object in question. Hence, the beautiful and the sublime involve appreciation of the aesthetic object for its own sake: 'both please for themselves' (§23, 5:244).

There are also important differences, however, which are significant to understanding the activity of imagination in both categories. Importantly, Kant's distinction between beautiful form and sublime formlessness identifies a different type of imaginative activity. Imagination and the understanding are in a harmonious free play in the beautiful, while the sublime calls forth an activity in which imagination and reason are less harmonious – the imagination being inadequate to the sublime, and reason adequate (§23; 5:244). The sources of the sublime response are linked to the physical properties of magnitude or power in nature but also to a failure of imagination. Imagination's activity, in contrast to the beautiful, is 'serious,' where some object is 'contra purposive for our power of judgment, unsuitable for our faculty of presentation, and as it were doing violence to our imagination, but is nevertheless judged all the more sublime for that' (§23, 5:245). Judgments of taste spring from the way in which the form of some object engages the cognitive powers in a free, harmonious play of the imagination and the understanding, where imagination is free from the

---

5 Kant, Immanuel (2000): *Critique of the Power of Judgment.* Paul Guyer, ed., Paul Guyer and Eric Matthews, trans. Cambridge: Cambridge University Press, [1790]. All page references are to this edition (and to the Academy edition as provided by the Cambridge edition). Hereafter abbreviated *CPJ*.

6 Kant uses *Erhabene* for 'the sublime' and *Schöne* for 'the beautiful'. The adjective, *erhaben*, means being raised, or elevated.

constraints of conceptualization. By contrast, judgments of the sublime are occasioned by natural objects in virtue of their appearing formless or unbounded, for example, 'shapeless mountain masses towering above one another in wild disorder with their pyramids of ice' (§26, 5:256).[7]

Accordingly, the activity of imagination is different; natural objects do 'violence to our imagination' by pushing it to the very limits of its powers, but nature is 'nevertheless judged all the more sublime for that' (§23, 5:245). In contrast to the tranquility of the beautiful, there is 'movement of the mind' and physical agitation in the sublime (§24, 5:247). In judgments of the beautiful, Kant claims that it is the free play of imagination and understanding which gives rise to a liking for the beautiful. In the sublime, imagination is conflicted: expanded yet also frustrated.

The impression of formlessness in natural objects is essential to Kant's attempts to distinguish judgments of the sublime and the complex activity of imagination. The pleasurable feeling of the beautiful arises from the free play of imagination and understanding in response to perceivable form, such as the graceful flight of a bird, and we have no difficulty in calling such objects beautiful. By contrast, the 'wild disorder' of 'shapeless mountain masses' means, that there is no particular or whole to perceive and grasp in imagination. So, in comparison to the beautiful it is more difficult to pick out an object of perception to call sublime. Kant's claim that we are mistaken if we call mountains sublime seems odd when considered in light of our phenomenological experience of mountains, but when observed in relation to his remarks on the beautiful, it becomes easier to understand. If formlessness is not graspable in a strict sense by the mind, it becomes very difficult to judge formless objects themselves as sublime.

That Kant is so careful to distinguish beautiful form from sublime formlessness shows the very indispensable role played by formlessness in our response. This response is essentially shaped by the way the disharmonious appearance of formlessness engages, and yet finally overwhelms, imagination. The aesthetic apprehension of this formlessness engages the mind in a particular way and gives rise to negative pleasure and awareness of the ideas of reason. So, even if ungraspable, the appearance of formlessness in a sublime object arguably plays just as important a role as form does in the beautiful object. Indeed, I be-

---

7 I have argued elsewhere for a nature-only interpretation of Kant's theory of the sublime. See Brady, Emily (2013): *The Sublime in Modern Philosophy: Aesthetics, Ethics, and Nature.* New York: Cambridge University Press. For other approaches supporting this interpretation, see Guyer, 1996; Uygar Abaci, Uygar (2008): "Kant's Justified Dismissal of Artistic Sublimity". In: *Journal of Aesthetics and Art Criticism,* 66:3, 237–51; Uygar, Abaci (2010):"Artistic Sublime Revisited: Reply to Robert Clewis". In: *Journal of Aesthetics and Art Criticism,* 68: 2, 170–173.

lieve we find a clearer phenomenology of the experience of sublime objects as compared to beautiful ones, given the vagueness so often associated with Kant's idea of the free play of imagination and the understanding.

At this stage, I should clarify that I interpret Kant's theory of the sublime as finding qualities in nature as well as humanity and its freedom to be sublime, rather than only the latter, despite his remarks concerning 'subreption' (§25, 5:257), when we substitute respect for our moral personhood with respect for nature.[8] On his view, it seems that sometimes we make a mistake and call nature sublime, when it is really the mind that is sublime. Kant's *a priori* approach, and his attempt to find a place for the sublime within his critical philosophy explain, at least more generally, why he puts strong – and very particular – emphasis on the mind and its powers as the true object of the sublime: 'the sublime in nature is only improperly so called, and should properly be ascribed only to the manner of thinking, or rather its foundation in human nature' (§30, 5:280).

However, there is sufficient evidence to recognize the causal role played by natural *objects* in judgments of the sublime – a point which is relevant to understanding imagination's activity as affected by formless phenomenal qualities. First, we find specific examples of natural objects and the qualities which cause sublime feeling, e.g., 'wide ocean, enraged by storms' (§23, 5:245); the Milky Way (§26, 5:256); 'mountain ranges towering to the heavens;' 'deep ravines and the raging torrents in them;' 'deeply shadowed wastelands inducing melancholy reflection' (General Remark, 5:269); 'starry heavens' (General Remark, 5:270). We can see various phenomenal qualities of these objects as well: 'raging', 'towering', 'wide'; and also, for example: chaos, 'wildest and most disruly order', devastation (§23, 5:246). Second, Kant discusses our reactions to formlessness, how imagination is overwhelmed, and how the mind is both attracted and repelled by sublime objects, giving a clear causal role for external nature in terms of how it affects the subject. We have, then, a range of objects and qualities affecting the mind in a particular way. Third, his accounts of sublimity use the language of experiencing phenomenal appearances (as we also see in the beautiful). For example, Kant writes that 'Nature is thus sublime in those of its appearances the intuition of which brings with them the idea of infinity' (§26, 5:255).

Imagination's activity in response to sublime objects is also significant for the complex affective response of the subject, a mix of pleasure and displeasure, or what Kant calls a 'negative pleasure.' This negative pleasure is described as a form of admiration or respect, and as such is not a strictly positive pleasure, as

---

8 Further support for this interpretation can be found in Brady, 2012, 91–109, and Brady, 2013.

we might find in the pleasure associated with the beautiful, which 'brings with it a promotion of life' (§23, 5:244). The mixed feeling of the sublime involves both attraction and repulsion, 'the feeling of a momentary inhibition of the vital powers and the immediately following and all the more powerful outpouring of them' (§23, 5:245), and the beautiful involves calm contemplation, while the sublime is marked by a 'movement' of the mind. Later, this movement is described as comparable to a 'vibration', 'a rapidly alternating repulsion from and attraction to one and the same object' (§27, 5:258).[9] Kant gives some indication of the nature of this feeling, which is reminiscent of earlier theories, especially those of Burke and Mendelssohn: 'astonishment bordering on terror, the horror and the awesome shudder....' (General Remark, 5:269). As imagination is overwhelmed, it is also expanded and elevated. In this way, imagination's distinctive activity can help us to track the complexity of sublime emotions, a mixture of both positive and negative feeling, though, overall, a positive feeling as we experience a sense of our own distinctive capacities.

3 The Expansion of Imagination and Aesthetic Freedom

Imagination became central to aesthetic theory in the eighteenth century and would continue to be important for Kant, Schiller, and the Romantic tradition in Britain and on the continent.[10] In many accounts, it is an essential mental activity in response to sublime qualities, an activity that, in turn, causes the distinctive mixed emotions which characterize this type of aesthetic experience. Imagination functions in ways that become significant for fashioning the character of the sublime response, for example, through expansion and acts of association. To understand the productive powers of imagination and their relationship to aesthetic freedom in the Kantian sublime, it will be useful to consider one key precursor and direct influence, Addison.[11]

---

9 See Guyer's discussion of the different and sometimes inconsistent ways Kant describes the relationship between pleasure and displeasure in the sublime (1996, 203–5; 208–9; 210–14). I follow Budd's interpretation of the sublime involving an emotional state with oscillation between two aspects, repulsion and attraction (*The Aesthetic Appreciation of Nature*. Oxford: Clarendon Press, 2002, 85).

10 For discussion of the role of imagination in precursors to Addison, see Karl Axelsson (2007): *The Sublime: Precursors and British Eighteenth-Century Conceptions*. Oxford: Peter Lang.

11 Gracyk notes that Addison's *The Spectator* appeared in German translation in 1745, and that the text is mentioned in Kant's pre-Critical work on the sublime in: Kant, Immanuel: "Ob-

Addison strongly influenced eighteenth-century aesthetic theory, and for our purposes here, his views are especially relevant for their emphasis on imaginative freedom. Imagination is expanded through attempting to present an image of the object to the mind:

> Our imagination loves to be filled with an object, or to grasp at anything that is too big for its capacity. We are flung into a pleasing astonishment at such unbounded views, and feel a delightful stillness and amazement in the soul at the apprehension of them. The mind of man naturally hates everything that looks like a restraint upon it, and is apt to fancy itself under a sort of confinement, when the sight is pent up in a narrow compass, and shortened on every side by the neighbourhood of walls or mountains. On the contrary, a spacious image of liberty, where the eye has room to range abroad, to expatiate at large on the immensity of its views, and to lose itself amidst the variety of objects that offer themselves to its observation.[12]

Freedom characterizes imagination in response to greatness in objects that are themselves unbounded and expansive. Imagination 'dislikes' being constrained, and the pleasure we feel is closely related to this freedom. In this passage, we also see Addison observing that we reflect on an image of our *own* freedom, as instantiated in experiences of the sublime in nature.

It is notable that it is the unconfined qualities of sublime *nature* which have this effect on imagination:

> [F]or though they [works of art] may sometimes appear as beautiful or strange, they can have nothing in them of that vastness and immensity, which afford so great an entertainment to the mind of the beholder.... The beauties of the most stately garden or palace lie in a narrow compass, the imagination immediately runs them over, and requires something else to gratify her; but, in the wide fields of nature, the sight wanders up and down without confinement, and is fed with an infinite variety of images, without any certain stint or number.[13]

In other passages, Addison contrasts the pleasant and designed beauty of gardens with the greatness of wilder landscapes. The contrast here is clear: between that which is ordered, designed and formal and that which is unbounded and disordered, in some sense. Also, the poet's imagination ought to have a grasp of nature, and in discussing *The Iliad*, Addison picks out how imagination is

---

servations on the feeling of the beautiful and sublime", trans. Paul Guyer, in Immanuel Kant (2007): *Anthropology, History and Education*, ed. Günter Zöller and Robert B. Loudon. Cambridge: Cambridge University Press, 23–62. See Gracyk, 1986, 217 n.; Crowther, 13.

12 Joseph Addison and Richard Steele, *The Spectator* (London, 1812), No. 412.

13 Addison and Steele, 1812, No. 414.

struck by 'a thousand savage prospects'.[14] We see here a kind of aesthetic freedom, where sublimity frees the imagination from constraint and enables us to enjoy images of our own freedom through aesthetic experience.

Addison's ideas are rooted in a broader notion running through aesthetic theory in the eighteenth century, brought forward mainly through the influence of Abbé Jean-Baptise Du Bos's *Critical Reflections on Poetry, Painting, and Music* (1719). Du Bos argued that the point and pleasure of art is to alleviate a bored, inactive mind and to stimulate the emotions.[15] Novelty, with variety, newness, and surprise, alleviates such boredom, but the sublime is also important in terms of the way it invigorates and expands imagination; the challenge that occupies imagination is essential to the distinctive mix of emotions felt in sublime experience.

Kant's theory of the sublime can also be understood as revealing a kind of aesthetic freedom much like we see in Addison, though it is situated within the contours of his own critical philosophy.[16] Although Kant gives new names to these categories, his distinction between the mathematically and dynamically sublime largely reflects one made by other theorists in the eighteenth century between sublimity connected to size, and that which is connected to power.[17] Interestingly, the distinction is drawn through imagination: 'the first is attributed to

---

14 Addison and Steele, 1812, No. 414.

15 See Burke, Edmund (1968): *A Philosophical Enquiry into the Origin of Our Ideas of the Sublime and the Beautiful*, ed. J.T. Boulton. Notre Dame: University of Notre Dame Press [Second Edition, 1759]. Editor's Introduction, p. lv, and Guyer, 2005, 16–17.

16 I should note that Baumgarten's aesthetic theory, and his ideas about imagination, also influenced Kant (see Guyer, 2005, 29), but I have discussed Addison here because we know he was a direct influence as well, and he has much more to say about imagination in relation to the sublime.

17 Crowther (1989, 107) cites a passage in which Addison appears to anticipate Kant's ideas on the mathematical sublime. 'The understanding . . . opens an infinite space on every side of us, but the imagination, after a few faint efforts, is immediately at a stand, and finds herself swallowed up in the immensity of the void that surrounds it; our reason can pursue a particle of matter through an infinite variety of divisions but the fancy soon loses sight of it . . . the object is too big for our capacity when we would comprehend the circumference of a world, and dwindles into nothing when we endeavor after the idea of an atom (Addison and Steele, 1812, No. 420). Kant's examples are also reminiscent of earlier accounts, and we find here his only mention of artifacts as sublime, specifically St Peter's in Rome and the pyramids in Egypt (examples offered by Kames and Gerard as well). We also find various discussions of architecture and ideas similar to the mathematically sublime – such as Burke's discussion of the 'artificial infinite' in relation to rotundas, temples and cathedral. See Burke, 1968, 74–75. Guyer cites the influence of Baumgarten, who outlined a distinction similar to the mathematically and dynamically sublime. See Guyer, Paul (2012): "The German Sublime After Kant". In: Costelloe, Timothy M. (Ed.): *The Sublime From Antiquity to the Present*. New York: Cambridge University Press.

the object as a *mathematical*, the second as a *dynamical* disposition of imagination' (§24, 5:247). Within the context of Kant's transcendental philosophy, these two forms of the sublime are not narrowly circumscribed aesthetic experiences. Rather, each in its own way puts us in touch with our power of reason, and reveals to us, through sensible experience, our capacity for freedom. In contrast to the beautiful, Kant writes that the sublime is 'related through the imagination...to the *faculty of desire*' (§24, 5:247).

In the mathematically sublime, the senses and imagination are pushed to the very limits of their powers when faced with the overwhelming size of natural objects such as high mountains or the night sky, which are suggestive of the infinite. Sublime size here is that which is absolutely or exceedingly great, that which is great 'beyond all comparison' (§25, 5:250). Instead of using a mathematical method of quantifying or measurement, the sublime involves an attempt to grasp a whole through a kind of 'aesthetic comprehension' executed by imagination. Through this attempt to aesthetically comprehend the absolutely great, and subsequent failure, 'the very inadequacy of our faculty for estimating the magnitude of the things of the sensible world awakens the feeling of a supersensible faculty in us' (§25, 5:250). As imagination fails to take in the sensible particulars of such vast magnitudes, in other words, we are made aware of reason's capacity to provide an *idea* of the infinite: 'Nature is thus sublime in those of its appearances the intuition of which brings with them the idea of its infinity' (§26, 5:255). The expansion and failure of imagination gives rise to inhibition, and an awareness of the power of reason and its capacity to take in the totality of the mathematically sublime. This leads Kant to a nominal definition of the sublime: 'That is sublime which even to be able to think of demonstrates a faculty of the mind that surpasses every measure of the senses' (§25, 5:257).

It is important to understand the particular way in which imagination is both strained *and* constructive in the mathematically sublime. The 'enlargement of the imagination' is satisfying *in itself*, and not merely as an essential step in the opening out of reason's power (§25, 5:248). It is expanded through striving toward ever-larger exhibitions of totality, with a feeling of straining toward the infinite while never reaching it. For despite its inability to intuit an absolute whole, it is through working toward that end that imagination demonstrates its ability to expand the limits of the graspable. Kant's architectural example of St Peter's basilica in Rome nicely illustrates the two aspects of imagination's activity, and indicates the feeling associated with expansion:

> For here there is a feeling of the inadequacy of his imagination for presenting the ideas of a whole, in which the imagination reaches its maximum and, in the effort to extend it, sinks back into itself, but is thereby transported into an enormously moving satisfaction. (§26, 5:252)

Clearly, imagination's role is generally more productive in the mathematically sublime than might first appear.[18] Not only do we see Kant articulating the active, expanded imagination as being satisfying in itself, but we can also see that its very inadequacy in representing ideas of infinity is an essential part – its vocation even – for bringing reason into play:

> But our imagination, even in its greatest effort with regard to the comprehension of a given object in a whole of intuition (hence the presentation of the idea of reason) that is demanded of it, demonstrates its limits and inadequacy, but at the same time its vocation for adequately realizing that idea as a law. (§27, 5:257)

To shed more light on this expansion, it is worth considering some similar ideas of expansion seen in imagination's generation of 'aesthetic ideas' through art.[19]

Kant's discussion of fine art identifies imagination as a productive and exhibitory or presentational power. In its non-reproductive function, imagination generates images and associations in relation to concepts, but in aesthetic judgments, imagination is unconstrained by conceptual thought, and, indeed, its particular role is to expand beyond the limitations of the phenomenal given. In our appreciation of fine art, imagination 'spreads its wings' and, through generating aesthetic ideas, exhibits a multitude of images or representations in relation to the actual images or symbols in art. An aesthetic idea is defined as:

> [T]hat representation of the imagination that occasions much thinking though without it being possible for any determinate thought, i.e., *concept*, to be adequate to it, which, consequently, no language fully attains or can make intelligible. (§49, 5:314)

Through its interaction with an artwork, imagination provides poetic or aesthetic insight into particular concepts, building on the inadequacy of literal expression or non-poetic language. In short, imagination fulfils the function of showing, where saying is inadequate.

These productive and exhibitory powers of imagination also function when the imagination strives to exhibit the totality of natural objects, and attempts to

---

18 Cf. Guyer, 2005, 158; Clewis, Robert R. (2009): *The Kantian Sublime and the Revelation of Freedom*. Cambridge: Cambridge University Press, 82–83.

19 I should note that in drawing on Kant's notion of aesthetic ideas in art to illuminate the role of imagination in the sublime, I am not thereby drawing any conclusions about the possibility of an artistic sublime in Kant's theory. For this sort of view in relation to aesthetic ideas, however, see: Wicks, Robert (1995): "Kant on Fine Art: Sublimity Shaped by Beauty".In: *Journal of Aesthetics and Art Criticism*, 53:2, 189–193; and Pillow, Kirk (2000): *Sublime Understanding: Aesthetic Reflection in Kant and Hegel*. Cambridge, MA: MIT Press; Crowther, 1989, 159 ff.

present, at the same time, all the units of magnitude that would come together as an absolute whole.[20] It succeeds to some extent, and in turn provides the essential launching pad for reason's role. Thus:

> [F]or the imagination, although it certainly finds nothing beyond the sensible to which it can attach itself, nevertheless feels itself to be unbounded precisely because of this elimination of the limits of sensibility; and that separation is thus a presentation of the infinite, which for that very reason can never be anything other than a merely negative presentation, which nevertheless still expands the soul. (General Remark, 5:274)

We can see that imagination's inadequacy in the face of reason is but one part of its activity in the sublime response, for it does much of the work in enabling us to grasp the idea of infinity through aesthetic rather than epistemic estimation. Although imagination is challenged to its limits, it 'acquires an enlargement and power which is greater than that which it sacrifices' (General Remark, 5:269). It is not solely an instrument of reason, but an essential activity in itself, making the very feeling of the sublime possible.

I would also argue that imagination's activity in the mathematical sublime amounts to an expression of *aesthetic* freedom, not unlike Addison's ideas, inducing a feeling of pleasure independent of any recognition our capacity for reason. The expansion of imagination brought on by the unbounded qualities of natural objects – vast deserts or the starry night – is enjoyed for itself, free from constraints, as it operates within aesthetic judgment. Although imagination functions in a more complex way compared to the 'free play' of the beautiful, it is more intense and invigorated in this context, with a freedom that appears to be larger or more extensive by comparison. Recognizing this feature of imagination enables us to reappraise the Kantian sublime as an aesthetic category very much in its own right.[21]

---

20 Gibbons also recognizes the exhibitory function of imagination in presenting aesthetic ideas and this same function in the mathematically sublime (1994, 150–151 n. 26). Although Kant's discussion of aesthetic ideas is focused on the arts, he seems to allow for aesthetic ideas to be generated also through aesthetic judgments of nature (*CPJ*, §51, 5:320).

21 It is also significant that freedom is not only a key aspect of aesthetic judgment and experience for Kant, but especially so with respect to natural objects. This can be gleaned from his emphasis on nature in both the beautiful and the sublime, and his association of free (rather than dependent) beauty principally with natural objects. More recently, Malcolm Budd and Ronald Hepburn have emphasized the particular freedom of aesthetic appreciation of nature in contrast to appreciation of artworks. See Budd, 2002, 106–109, and Hepburn, Ronald W. (1984): "Contemporary Aesthetics and the Neglect of Natural Beauty". In: *Wonder and Other Essays*. Edinburgh: Edinburgh University Press, 9–35; first published 1966 in: Williams, Bernard and

4 Imagining Threats

Imagination's positive activity is not limited to the mathematically sublime, although its 'expansion' is most explicit there. For in the dynamically sublime, the power of natural objects 'raises imagination to the point of presenting those cases in which the mind can make palpable to itself the sublimity of its own vocation even over nature' (§28, 5:262). What does Kant mean by this?

The dynamically sublime is mainly concerned with nature's power, and Kant's examples are very much in keeping with stock examples offered by earlier theorists:

> Bold, overhanging, as it were threatening cliffs, thunder clouds towering up into the heavens, bringing with them flashes of lightning and crashes of thunder, volcanoes with their all-destroying violence, hurricanes with the devastation they leave behind, the boundless ocean set into a rage, a lofty waterfall on a mighty river, etc., make our capacity to resist into an insignificant trifle in comparison with their power.(§28, 5:261)[22]

As in the mathematically sublime, imagination is overwhelmed, yet here it is power, and not only size, which overwhelms, and it is explicitly practical reason that is said to be 'awakened.' In relation to these fearful, powerful forces we are made to feel physically small. This feeling is possible only under the conditions of being situated in a safe place, and not seriously threatened. Kant's remarks on the disinterestedness of judgments of the sublime are consistent with his claims about judgments of taste, which are independent of interests in the 'real existence' of an object (we take an aesthetic rather than, for example, utilitarian interest in the object; §2, 5:205). More specifically, disinterestedness is articulated through an appreciation that cannot be distracted by concerns over one's own safety:

> Someone who is afraid can no more judge about the sublime in nature than someone who is in the grip of inclination and appetite can judge about the beautiful. The former flees from the sight of an object that instills alarm in him, and it is impossible to find satisfaction in a terror that is seriously intended. (§28, 5:261)

Alan Montefiore (Ed.). In: *British Analytical Philosophy*. London: Routledge and Kegan Paul, 285–310.

22 Clewis points out that Kant's list includes objects that can be both mathematically and dynamically sublime depending on which feature has more relevance in any particular experience (2009, 67).

Sublime objects are correctly described as fearsome, where actual fear is not realized because of aesthetic distance and the feeling of our capacity to overcome nature's might. The line between actual fear of nature and the sublime response is by no means sharp, but Kant is certainly able to convey the distinctiveness of this type of appreciation.

If we are in fact safe from 'deep ravines and the raging torrents in them', how is it possible to experience the anxiety mixed with pleasure characteristic of sublime feeling? What, then, is there to be anxious of, if in fact we are safe? The answer, I believe, is that we imagine an outcome in which we are threatened. As Kant puts it, 'the astonishment bordering on terror...is, in view of the safety in which he knows himself to be, not actual fear, but only an attempt to involve ourselves in it by means of imagination...' (General Remark, 5:269). In this way, imagination plays a fundamental part in facilitating sublime feeling through imagining – visualizing or a more embodied imagining – the physical harm involved in, say, tumbling into a deep ravine, which *might* occur if we were not in a secure place. Such imaginings will, importantly, help to elicit the same kinds of feelings as if we were tumbling down. Consider another case. We can become excited by the force of a thunderstorm from a distance, or perhaps while standing at the edge of it, but not so easily if deeply concerned about being struck by lightning – though we can certainly imagine this possibility, caught out in the middle of a field, say. It seems that the sublime requires us to strike a balance between being completely overwhelmed or completely underwhelmed by these tricky situations. Disinterestedness, imagination, and a position of safety together facilitate this balance.

This point helps to reveal, again, how important imagination's activity is to sublime feeling. Imagining the kinds of things that could happen to us amidst nature's forces will sometimes, though perhaps not always, underpin our quasi-fear and astonishment (it may be the case that the mere perception of powerful qualities causes these feelings).[23] Imagination and feeling closely interact in the sublime response; while judgments of the beautiful also involve this close interaction, as the free play of imagination and the understanding gives way to pleasure, in the sublime this interaction is more complex. Although it is ambiguous whether the displeasure and pleasure of the sublime are felt simultaneously or oscillate, the source of displeasure is the shocking effect on imag-

---

23 Budd underplays this particular function of imagination in the dynamically sublime, and although I agree that Kant does not make it a necessary feature, it certainly plays some role. See Budd, Malcolm (1998):"Delight in the Natural World: Kant on the Aesthetic Appreciation of Nature. Part III: The Sublime in Nature".In: *British Journal of Aesthetics*, 38:3, 242–244 (reprinted in Budd, 2002)

ination brought about by the formlessness of natural objects. Thus, we find that imagination's activity is essential to the distinctive character of dynamically sublime affect.

We also discover a function of imagination that is different from its activity of freedom and expansion in the mathematically sublime. In imagining threatening outcomes, imagination is given content through the different ways we imagine nature to be threatening, depending on the particular forces at work – a deep ravine, thunderstorm, rocks falling down from high above, etc. In the dynamically sublime, imagination seems to be, at least sometimes, operating projectively. That is, we imaginatively project ourselves falling into the ravine – perhaps not intentionally, but because we cannot help but to be drawn into the great depths of the place in that sort of way.

5 Imagination, Association, and Identification

The activity of imagination in the dynamically sublime suggests two additional ways it is given content through engagement with natural objects and their qualities: association and identification. First, it is notable that some eighteenth-century theories of the sublime propose that imagination is also engaged in making associations between some aesthetic object and some relevant cultural event. For example, Archibald Alison holds that: 'imagination is seized, and our fancy busied in the pursuit of all those trains of thought which are allied to this character of expression'.[24] The aesthetic value and sublimity of landscapes can be increased through the presence of various cultural elements, and by bringing together natural landscapes with cultural events or poetic imagery. Thus 'plain scenes' are transformed through associations with battles having taken place there, and the Alps become more sublime through associations with Hannibal's March over them. Likewise, a deep chasm has an even greater effect when associated with the local story of a woman flinging herself into it.[25]

Is there any scope for imaginative associations in Kant's theory? As one influence on Kant, Mendelssohn does not explicitly assign a role to imagination beyond what sounds like an activity connected to the mathematical sublime, where the multiplicity of something 'prevents all satiation, giving wings to imag-

---

24 Archibald Alison, *Essays on the Nature and Principles of Taste,* in Francis, Lord Jeffrey, *Essays on Beauty,* and Archibald Alison, *Essays on the Nature and Principles of Taste,* 5th ed. (London: Alexander Murray, 1871; reprint by Kessinger Publishing), 69.
25 Alison (1871), 76–77.

ination to press further and further without stopping',[26] and a broader recognition of imagination's role in the artistic genius responsible for sublime works of art. For Mendelssohn it appears to be the senses, mainly, that are overwhelmed, and he is critical of the role of associative imagination, because he sees the novel and unexpected as central to the feeling of the sublime, that is, 'astonishment'. Sublime feeling arising through association suggests a less immediate type of effect; additionally, it is also the case that associations do not arise from the sublime itself, since Mendelssohn takes the impact of the sublime on the mind to be too strong to produce anything but a state of stupor, 'an absence of consciousness'.[27] These remarks point to an *emptying* of the mind in the reaction of astonishment, rather than filling it up with a train of associations.

Although Kant distinguishes an important role for imagination, it is unlikely that he would want to extend it to Alison's associationism, at least because this would make the foundations of sublime judgment too empirical and diverse, undermining universal communicability.[28] Kant is careful to distinguish his aesthetic theory from empirically-oriented ones with which he was familiar, such as those of Hume and Burke (see, e.g., General remark, 5:277–278). For Kant, judgments of the sublime are grounded in formless qualities in objects, the overwhelming of the senses and imagination, and the feeling of being a measure to nature, rather than historical events or other cultural information.[29] His focus on the natural sublime, unlike the attention paid to sublimity in art, literature and poetry as well as nature, by earlier theorists, also suggests a reason for his lack of interest in cultural associations to sublime nature. This focus is not accidental, I believe, given the broad interest in nature and human nature in the *Critique of the Power of Judgement* and in his moral philosophy.

Nevertheless, some of his remarks suggest the possibility of a role for something like associations in the sublime response. Kant insists that the pure judgment of the sublime is concerned not with objects that 'presuppose the concept of an end' but rather with the free imaginative engagement which characterizes aesthetic judgment. For example:

---

26 Mendelssohn, Moses (1758): "On the Sublime and the Naïve in the Fine Sciences".In: Dahlstrom, Daniel O. (Ed., trans.): *Philosophical Writings*. Cambridge: Cambridge University Press, 1997, 192–232, 195.

27 Mendelssohn, Moses, 1997, 192–232, 199.

28 Guyer points out that Alison was unaware of Kant's *CPJ*, and he emphasizes the empirical and *a priori* contrast in their approaches. See Guyer, 2005, 209.

29 See, also, Kant's remarks about the sublime having its foundation in human nature rather than culture (§29, 5:265).

> [W]e must not take sight of the ocean as we *think* it, enriched with all sorts of knowledge...-
> for example as a wide realm of water creatures, as the great storehouse of water for the
> evaporation which impregnates the air... for this would yield teleological judgments; rather,
> one must consider the ocean merely as the poets do, in accordance with what appearance
> shows, for instance, when it is considered in periods of calm, as a clear watery mirror
> bounded only by the heavens, but also when it is turbulent, an abyss threatening to devour
> everything, and yet still be able to find it sublime (General Remark, 5:270).

This particular example characterizes the ocean in both its mathematically and
dynamically sublime expressions. In an example just before this one, Kant
makes a similar point with regard to the 'starry heavens' which we must take
'as we see it, merely as a broad, all-embracing vault' (General Remark, 5:270).
In urging that our aesthetic judgments do not have a basis in facts or empirical
concerns, he may be pointing to the production of aesthetic ideas generated
through the activity of the productive imagination (here in relation to nature
rather than art). Kant writes that aesthetic ideas function to 'animate the mind
by opening up for it the prospect of an immeasurable field of related representations' (§49, 5:315). Now, these images will be less concrete than those proposed
by Alison, but they *are* a type of sensible image for Kant. If we accept this point –
or some reconstruction of Kant's views along these lines – then we have a form
of productive imaginative activity which is given content through engagement
with the qualities of some aesthetic object.[30]

Secondly, we also find that the expansive activity of imagination can itself,
sometimes, be given content through *identification* with sublime objects. How
might this take shape? It could be that the expanded nature of imagination's activity reflects the very expansive qualities of the natural objects with which it engages, suggesting the possibility of a kind of identification between imagination
and sublime qualities. As far back as Longinus, a view running through several
theories of the sublime is that as the imagination (or more generally, the mind) is
expanded we also experience a sense of our own capacity to take in vastness or
great power, thereby evoking a sense of our own powers. John Baillie traces the
casual chain in the sublime response, explaining how sublimity becomes reflected in the subject: 'vast objects occasion vast sensations, and vast sensations give
the mind a higher idea of her own powers...'.[31] He rightly emphasizes that without the sensations of the external sublime, we would never experience the ex-

---

30 I would argue that such sensible images will not prevent any *new* experience of the same
thing – e. g., the Grand Canyon – from being judged sublime again, since on Kant's account the
immediate, sheer overwhelming feeling will likely push into the background previous experiences and accompanying images.

31 Baillie, John (1747): *An Essay on the Sublime.* London, sect. I, p. 7.

pansion that raises this awareness of our own admirable capacities. We can see the positive aspect of the sublime operating here, with some new appreciation of the self accompanying appreciation of external qualities. In line with this idea, Gerard's remarks are strikingly similar to Kant's:

> We always contemplate objects and ideas with a disposition similar to their nature. When a large object is presented, the mind expands itself to the extent of that object, and is filled with one grand sensation, which totally possessing it, composes it into a solemn sedateness, and strikes it with a deep silent wonder and admiration: it finds such a difficulty in spreading itself to the dimensions of its object, as enlivens and invigorates its frame: and having overcome the opposition which this occasions, it sometimes imagines itself present in every part of the scene, which it contemplates; and, from the sense of this immensity, feels a noble pride, and entertains a lofty conception of its own capacity'.[32]

These points about the expansion and invigoration of the mind, and a sense of one's own loftiness, would appear to anticipate Kant's ideas.[33] However, Gerard also captures how projection becomes a kind of identification with the boundlessness which one finds in external nature. It's not clear that Kant ever conveys the idea of the mind imagining itself 'in every part of the scene', but there is perhaps some version of this in his view that we measure ourselves in relation to nature which is great and powerful, e. g., a thunderstorm. We are made to feel physically small when positioning our puny, phenomenal selves in comparison to physical nature that is greater than us. But in this comparative act, we also discover that we do have the capacity to take in and become a measure to nature's greatness – to share in its sublimity. So, when we measure ourselves in this way, there is something akin to an identification with nature's sublimity (even though we discover that we are, ultimately, independent from it, as revealed through our capacity for reason):

> 'Thus nature is here called sublime merely because it raises the imagination to the point of presenting those cases in which the mind can make palpable to itself the sublimity of its own vocation even over nature'. (§28, 5:262)

---

32 Gerard, Alexander (1759): *An Essay on Taste*. London, I.2., 168. Cf. Rachel Zuckert's discussion of Gerard in, 'The Associative Sublime: Gerard, Kames, Alison and Stewart', in Costelloe, 2012.

33 Many thinkers on the sublime shared Gerard's and Baillie's ideas, but Gerard is especially relevant for this particular point about identification and, also, because we know that Kant was familiar with Gerard's work. See Guyer,Paul (2011): "Gerard and Kant: Influence and Opposition".In: *Journal of Scottish Philosophy*, 9.1, 59 – 93.

The result of this imaginative comparison is that we feel great – elevated – much like the greatness we find in nature, except that our greatness is found in our capacity for reason and freedom, rather than in any sensible aspects of human nature:

> [W]e gladly call these objects sublime because they elevate the strength of our soul above its usual level, and allow us to discover within ourselves a capacity for resistance of quite another kind, which gives us the courage to measure ourselves against the apparent all-powerfulness of nature. (§28, 5:262)

We respond with admiration for nature's power, but also for ourselves as having powers independent of sensible nature. So, the act of comparing and, then, finding ourselves to be sublime, *like* but also *unlike* nature, could be interpreted as an expression of partial imaginative identification.

6 Imagination and Moral Freedom

In section III, I discussed how imagination's activity of being expanded through engagement with sublime vastness in nature enables the subject to experience a kind of aesthetic freedom, echoing Addison's early discussion. In the dynamically sublime, however, we find that imagination's role is extended beyond aesthetic freedom, as it anticipates a feeling for moral freedom. Here, the overpowered imagination and subsequent opening out of reason provides us with a feeling of freedom through aesthetic rather than moral judgment and experience. Kant ultimately connects this to a feeling of respect or admiration for our moral vocation, that is, 'the idea of humanity in our subject' (§27, 5:257). In the dynamical mode of the sublime, we gain an awareness of negative practical freedom, or freedom from the senses which necessarily determine us, that is, freedom from our *internal* nature.[34]

As I noted earlier, my interpretation of the Kantian sublime places it firmly in the aesthetic domain. Although there are key links to be made to the notion of practical reason, the foundation of Kantian morality, it is important to empha-

---

34 Negative freedom for Kant is independence from determination by our internal, sensible nature. As Clewis puts it, through judgments of the sublime, we grasp that 'the natural object has no dominion over us, in the sense that we have the ability not to be determined by inner nature and natural impulses' (Clewis, 2009, 68). In Kant's moral philosophy, negative practical freedom can be contrasted with positive practical freedom where, in the latter, we are self-determining, acting from duty.

size that judgments of the sublime, like judgments of taste, only *prepare* us for morality, rather than actually placing us in the situation of an agent making a moral judgment.[35] In the 'General Remark,' Kant thus states only that the sublime response is 'a disposition of the mind that is similar to the moral disposition' (5:268). Moreover, it would be a mistake to interpret the sublime response as involving an intellectual recognition – or actual understanding – of our moral freedom. Instead, the sublime is firmly placed within the realm of feeling and, as a quality of *aesthetic* experience, gives a felt sense of our moral capacities only.[36] Just what that feeling amounts to as a kind of awareness rather than understanding is not always clear, but Kant is consistent in the language he uses to describe how ideas of reason are: 'awakened in us' (General Remark, 268); 'a feeling that we have self-sufficient reason' (§27, 5:258); and how the sublime 'arouses a feeling of the supersensible' (§27, 5:258). It is helpful to keep in mind the wider project of the third Critique, which considers human actions as independent from nature yet also effective within that realm. Aesthetic experience and judgment provide us with a sensuous, felt grasp of morality and our potential in that realm.

These points provide evidence, then, for understanding how imagination is functioning as part of the exercise of aesthetic judgment, and in this context, operating in a mode when it is the least constrained. Again, this activity is crucial, for without imagination's being shocked, overpowered and facilitating the fear of the dynamically sublime, we would not enjoy a sense of ourselves as morally independent from our inadequate, phenomenal selves.

7 Conclusion

Imagination has been shown to be a complex and expansive mental power which is essential to the distinctive feeling of the sublime response. In being challenged by sublime qualities of vastness or power in nature, imagination is overwhelmed, yet at the same time, expanded, invigorated, and elevated. Rather

---

35 See *CPJ*, §59 for Kant's discussion of beauty as a 'symbol of morality'. In his General Remark (5:267), Kant sets up this contrast: 'The beautiful prepares us to love something, even nature, without interest; the sublime, to esteem it, even contrary to our (sensible) interest'.
36 Matthews provides support for the importance of feeling rather than conceptual judgment as the basis of the sublime (1996, 177–8). In *Kant and the Experience of Freedom* (214–15), Guyer supports Kant's claim that that it is imagination alone which is at work in the sublime response, but in *Values of Beauty* (2005, 156–61), in the context of judgments of ugliness, he contends that judgments of the sublime do involve a *recognition* of our power of reason even if this is indicated through feeling.

than playing the role of a weak companion to reason, imagination reaches toward a presentation of seemingly infinite things and enables us to compare ourselves to the might of stormy seas and the like, facilitating a feeling for capacities which lie beyond our sensible nature. We have found that it functions in different ways towards these ends, including projection, identification, and, in some sense, association. Its activity involves a feeling of freedom which brings about pleasure – not only a form of aesthetic freedom, that is, an unconstrained imagination expanded in response to sublime natural qualities, but also a feeling for the freedom characteristic of moral judgments. In all of these positive ways, imagination drives the sublime and defines the very complex character of this type of aesthetic experience.

Martin Schönfeld
Imagination, Progress and Evolution

1 Scholarship

What role does the imagination play? What is the power that performs this role? And what are its roots? The best place for answers to these questions is in the Critiques. The power of imagination (*Einbildungskraft*) appears in the aesthetic epistemology of the *Critique of Judgment*. There it functions as a constitutive aspect of reflective judgments and works as a mediator between the realms of knowledge and desire. It also appears in the general epistemology of the *Critique of Pure Reason*. There it mediates between the two stems of knowledge, sensibility and understanding, an activity that yields temporal schemata to link empirical intuitions to pure concepts. Most importantly, imagination is the power that the keeps mind unified; it creates the synthetic unity of apprehension to anchor cognitions in the subject.

The power of imagination emerges overall as a dynamic agency in the critical architectonic. It is less of a structural element and more of a clamping device; less brick, and more mortar: it holds things together, links data flows, and mediates between cognitive structures. And this seems to be about all that can be determined, for a scholarly inquiry into Kant's concept of the imagination quickly hits limits. The account of the imagination in the first and last of the Critiques is neither very extensive nor particularly crisp. Kant has far less to say about this power than about its *effects:* the schematism, reflective judgment, or the synthetic unity of apperception. This is odd. Imagination is so crucial that without it everything would fall apart, yet little is said about it. It is so basic that it hides in the basement.

Sometimes what lurks in the root cellar is a monster. As a constitutive element in reflective judgments, the power of imagination is a necessary condition for the cognition *of beauty.* As the interface between stems of knowledge, it serves as a condition *of cognition as such.* And as the source of unified apprehension, it is a condition *of consciousness as such.* These are implications so enormous, so uncanny, that they are unsettling. What are we dealing with here? Where does a pursuit of the imagination lead to?

Then again: how should we go about this pursuit? To launch into an inquiry, we first need to decide on the method. And right at the beginning, inquiring into the imagination hits a fork in the road. One path is that of scholarly analysis. It proceeds as a descriptive reconstruction of Kant's use of the imagination, as an exclusive concern with what Kant actually wrote. Another path is that of conceptual synthesis. It proceeds, by comparison, as a *prescriptive* construction, of a Kantian use of the imagination; that is to say, as a critical, creative, and contextual concern with what Kant, considering what he wrote elsewhere, *should* have written about the imagination. Let us consider the scholarly path first. It begins with the gathering of textual evidence and continues with the study of the collected material. This path is broad, safe, and well-trodden. Three rules mark it out: observation, preservation, and obedience.

Observation: the first rule is to *proceed empirically.* Any statement on the imagination follows from scrutiny of the material. First we read the text; next we take stock and report. This is scholarship. 'Observation' means the physical act of focusing on the words and the cognitive attitude of submitting to their information. The text guides analysis. The inverse holds too: textual limits set analytic boundaries. There is nothing to report when there is nothing to be read. Better be silent then.

Preservation: the second rule is: *handle with care.* Texts demand gentleness from the scholar, all the more so when they are artifacts in the history of ideas and thus relics from the past. The scholar is like an archeologist who unearths a sculpture to clean and preserve it. The end of analysis is to reveal structure, to 'clean it,' and to make it intelligible, to 'preserve it'. Commentary subsequent to analysis serves merely an explicative purpose. This is not to insinuate an interpretive principle of charity as the *sine qua non* of scholarly analysis, since such analysis is perfectly compatible with a critical examination. But even in its critical variants, scholarship deserves to be called scholarship only if it is firmly grounded in analytic scrutiny of the texts. Scrutiny, in archeology or scholarship, must not change the object.

Obedience: the third rule is that *the text is always right.* Just as we cannot alter the text, we cannot read anything into it from the outside either. Of course, in actual hermeneutic practice, exegesis without interpretation (and interpretation without manipulation) hardly ever happens. It is unrealistic to assume that one can read a text without bringing something to the text in the sheer act of reading. Nonetheless, scholarly analysis aspires to keep hermeneutical manipulation and interpretive projection to a minimum. From a scholarly vantage point, manipulation, while unavoidable, is not particularly desirable, and projection, if not kept to a minimum, simply reveals sloppy research habits. Projecting a creative idea skews analysis: what one takes from the text would be what one inserted in the first place.

Textual obedience matters especially when analysis confronts holes. Should a gap open up in the textual information, interpretation may fill the lacuna merely by what can be extrapolated from the edges. Patching up holes or adding on replacements can be done in archeology and scholarship in minimally invasive ways, as simple, linear extensions from data. If an amphora has a hole in the belly or lacks a side handle, the archeologist can patch the hole or glue on a new handle according to the guidance by the vase. A patch fills the gap by stretching surface right across; a handle covers a stump to mirror the other handle. The object demands obedience even in partial absence. Deviation from data leads astray, especially when the material tells archeologists or scholars the opposite of what art lovers or creative readers hope to add.

2 The Moral Litmus Test

How would the rules of scholarship guide our inquiry into the imagination? Consider the riddle of the imagination in moral cognition. Assuming an action is right if its maxim can serve as a principle of legislation (4:402.7–9), how can we find out whether an intended action fits the bill? Evidently one must transform, in the mind, the maxim at issue into a principle and imagine what doing so would entail. The mental transformation, from single act towards its replication as a collective pattern, shows the maxim's value.

The value is positive if replication reveals an open lane, without roadblocks coming into view to slow down and stop replication. It is negative if the envisioned replication spawns a destructive iteration spiraling into oblivion. Positive value means the intended deed fits the bill, and we can go ahead with the considered plan of action; negative value means the maxim fails to universalize and is morally wrong.

Determining the value of the maxim thus turns on a method both virtual and quasi-Darwinian. The determination of the value happens *prior* to the enactment of the maxim, and the decision hinges on the reproductive fate of the maxim in a social environment. If a scenario suggests itself in which the offspring prevails, the maxim passes the test; if the imagined offspring can be seen to perish instead, the maxim will fail. This is the moral litmus test. Without it, the Categorical Imperative would make little sense. The test is fundamental to Kantian ethics.

Elsewhere in the Critical System, the power of imagination appears to be equally fundamental. Without the power in the basement, there would be no aesthetic cognition and no general cognition either. Is the capacity to determine the value of maxims—moral cognition—possible without the imagination? More precisely, does the moral litmus test require the power of the imagination at work, or does it not? The answer is clear, or so one would think, *and yet Kant happens to reject it:* hence the riddle.

So one needs to proceed cautiously and ask again: is the power of the imagination needed in moral cognition? Popular examples of replication-failure, and a starting point for an answer, would be the opposites of the Fifth, Seventh, and Eighth Commandments (on killing, stealing, and lying). If thou shalt kill was universal law, the prospect of all killing all would be inescapable; if that went on, everybody would kill until there was no one left, thus ending the replication. Similarly, if thou shalt steal were law, and everyone stole everything, replication would spiral into oblivion. Possessions would circulate among thieves, subverting the idea of property and reducing theft to absurdity. Nothing can be stolen if everything is loot. Lastly, if thou shalt lie, and everybody did, communication would collapse, and lying would stop. In each case, replication is self-reducing, revealing that killing, stealing, and lying are the wrong thing to do.

There are philosophical difficulties in the litmus test. Obviously, the test fails as a means to identify the right scope of maxims under consideration, most famously lying: does the negative value mean lying is wrong even when the act would mislead an inquiring murderer (8:426.14–427.26)? Still, the test discloses a difference in the potential of practices. The test anticipates what we call today *sustainability*—the highest policy principle of civilization and the unifying strategy during the climate crisis. While there is an overlap of sustainable and moral actions, their ranges are not coextensive: some practices can be replicated indefinitely and *still* be bad (think child abuse, factory farming, or patriarchy); others can end when replicated and nonetheless be good (think theft—global transition to sustainability will require the radical curtailing of corporate and private property). Such difficulties are limits rather than flaws: not all of the good comes into view with universalizability, but some of it does. This limitation, of course, raises the question of what other means, if any, we do have to broker the difference.

But to return: what is the role of the imagination here? Is it not the case that determining the value of a maxim by means of the test requires us *to imagine* what would happen if the maxim were to advance to a law? If so, then we might infer that imagination plays a pivotal role not only in general and aesthetic cognition, but in moral cognition as well. Kant describes the procedure of the litmus test in *Groundwork for a Metaphysics of Morals*, when discussing whether I can make a false promise when in distress and without intention of keeping it. He explains:

> However, to inform myself in the shortest and yet infallible way about the answer to this problem, whether a lying promise is in conformity with duty, I ask myself: would I indeed be content that my maxim (to get myself out of difficulties by a false promise) should hold as a universal law (for myself as well as for others)? And could I indeed say to myself that every one may make a false promise when he finds himself in a difficulty he can get out of in no other way? Then I soon become aware that I could indeed will the lie, but by no

means a universal law to lie; for in accordance with such a law there would properly be no promises at all, since it would be futile to avow my will with regard to my future actions to others who would not believe this avowal or, if they rashly did so, would pay me back in like coin; and thus my maxim, as soon as it were made a universal law, would have to destroy itself. (4:403.2–17; tr. M. J. Gregor 1996: 57)

Answering the question of the value of the maxim requires posing another question first (of whether I can will the maxim as a law). Answering the second question yields the answer to the first. The heuristic process involves steps, each being a stride from one cognitive location to the next. In the first step, the maxim is singled out, countenanced as a law, and queried on its potential: 'would I [...] be content that my maxim [...] should hold as a universal law?' In the second step, this query is pursued by envisioning a situation that brings consequences to light: suppose I did make a false promise, what might happen to me? Likely is the loss of my credibility ('others who would not believe this avowal'); dangerous is the risk of revenge ('pay me back in like coin'). In the third step, the gathered results radiate across the collective. Suppose the potential was universally actualized, suppose everyone acted that way, what would likely happen to the collective? The replication lets consequences be seen all the way to the end (eventually there 'would properly be no promises at all'). The fourth and final step is the cognitive pull-back from the collective chain to its self-reductive pattern. The self-reduction is seen ('as soon as it were made a universal law, [it] would have to destroy itself') and finally rated: 'I could indeed will the lie, but by no means a universal law to lie'. Hence lying is morally impermissible.

In the *Critique of Practical Reason*, Kant writes:

The rule of judgments under laws of pure practical reason is this: ask yourself whether, if the action you propose were to take place by a law of the nature of which you were yourself a part, you could indeed regard it as possible through your will. Everyone does, in fact, appraise actions as morally good or evil by this rule. Thus one says: if *everyone* permitted himself to deceive hen he believed it to be to his advantage ... and if you belonged to such an order of things, would you be in it with the assent of your will? Now everyone knows very well that if he permits himself to deceive secretly it does not follow that everyone else does so ... accordingly, this observation of the maxim of his actions with a universal law of nature is also not the determining ground of his will. Such a law is, nevertheless, a *type* for the appraisal of maxims in accordance with moral principles. If the maxim of the action is not so constituted that it can stand the test as to the form of a law of nature in general, then it is morally impossible. (5:69.20–70.1; tr. M. J. Gregor 1996: 196)

This second account enlarges the first: there is a process, and it is the same for all; everyone appraises actions by this rule. According to the *Critique of Pure Reason*, the imagination yields patterns, schemata. Here, in the *Critique of Practical*

Reason, the pattern is called 'type,' not schema (*Typik*; 5:69.36, 69.19). Is this not an interpretive gap that can be filled by stretching across a surface (the imagination at work) from edges elsewhere (the other two Critiques)? But Kant dispels this impression. A type is not a schema; and no, the imagination is not at work; the understanding is. He explains why:

> To a natural law, as a law to which objects of sensible intuition as such are subject, there must correspond a schema, that is, a universal procedure of the imagination (by which it presents a priori to the senses the pure concept of the understanding which the law determines). But no intuition can be put under the law of freedom (as that of causality not sensibly conditioned)—and hence under the concept of the unconditioned good as well—and hence no schema on behalf of its application *in concreto*. Thus the moral law has no cognitive faculty other than the understanding (not the imagination) by means of which it can be applied to objects of nature, and what the understanding can put under an idea of reason is not a *schema* of sensibility but a law, such a law, however, as can be presented *in concreto* in objects of the senses and hence as a law of nature, though only as to its form; this law is what the understanding can put under an idea of reason on behalf of judgment, and we can, accordingly, call it the *type* of moral law. (5:69.5 – 19; tr. M. J. Gregor 1996: 195 – 196)

The obedience-rule of scholarly analysis forbids inquiry to stray beyond text. The imagination is at work in aesthetic cognition and in general cognition, but not in moral cognition. Despite looks to the contrary, the moral litmus test is performed without resorting to the power of imagination. Why? Kant says so. He gives a cogent reason, in line with other claims of the critical system: the moral law does not belong to the sensible world; the imagination, however, does; so the imagination cannot yield the pattern of the moral law. Since the moral law belongs to the intelligible world, to which the understanding has access, it is the understanding that can yield the pattern: a type, not a schema.

For the scholar this settles it: when performing the moral litmus test, Kant states that imagination is not used. There may be sound interpretive reasons for reading the power of the imagination into the moral realm (it would be better to stick to one cognitive power throughout and we've already settled on the imagination; it sounds ad hoc to give the understanding more power than initially laid out; introspection reveals that imagination is at work in the moral litmus test; and so on), but a corollary of the obedience-rule, the projection-taboo, rules them out. It does not matter what critical readers want, regardless how good their reasons may be, it matters what the writer says, and in this case he says no.

In sum, the rules of good scholarship terminate the inquiry. Scholars of the imagination can read the *Critique of Pure Reason* and the *Critique of Judgment*, but there is nothing for them to read elsewhere in the oeuvre. And anyone who looks for the imagination in the *Groundwork*—or in *Perpetual Peace* and *Universal Natural History*—is not a scholar. At this precise boundary Kant scholarship comes to an end.

3 Geisterbeschwörung

This does not mean the inquiry into the imagination ends, for the choice between empirical-analytic scholarship and no inquiry at all is a false dichotomy. There is another option, conceptual synthesis, the other path from the fork encountered in the beginning. Here, philosophy relies on analysis and observation without being tied to them. Rational creation of insight circumvents the boundary.

At first glance, this seems trivial. Philosophy is more than scholarship, and reasoning is more than analysis. Still, conceptual synthesis is not all that conventional. On that path, observation comes first, reflection later. Dissecting data belongs to it as much as putting the disassembled factoids back together. Putting-together, here, is not reconstruction of items back to the original aggregate. Conceptual synthesis puts pieces back together into a whole, which is more than the sum of their parts, because it is a configuration with odd angles and tensions that spark heuristic interplays and braid new meaning. Conceptual synthesis is not only an alternative to scholarship but also its creative consequence.

A common alternative to scholarship, with regard to classical texts, is critique. It is on display in analytic philosophy (e. g. Jonathan Bennett's *Kant's Analytic* and *Kant's Dialectic*) and in postmodern deconstruction (think of the passages on Kant in Richard Rorty's *Philosophy and the Mirror of Nature*).[1] In either mode, a thinker is taken to task, examined, and gauged with yardsticks foreign to the thinker's concerns. Unsurprisingly, the thinker so estimated usually does not measure up. Critique can lead to understanding, but in these modes, it remains sophistry. Such exercises in skepticism are self-serving, ideological, and unproductive. Conceptual synthesis aspires to be different. It engages with classical texts not to make a point but to extend understanding, and thus proceeds on a path distinct from both scholarship and criticism.

A trailblazer of this distinct approach was Heidegger. A useful illustration is his 1927 seminar *Basic Problems of Phenomenology*. When discussing a thesis by

---

1 Bennett, Jonathan (1966): *Kant's Analytic*. Cambridge: Cambridge University Press, 138; Bennett, Jonathan (1974): *Kant's Dialectic*. Cambridge: Cambridge University Press, 288; Rorty, Richard (1979): *Philosophy and the Mirror of Nature*. Princeton: Princeton University Press, 392–394.

Kant (that being is not a predicate), he encounters an interpretive impasse, whose upshot he finds unsatisfactory. He writes:

> Should we end on this critical note? A criticism, which is merely negative, and which fails to be constructive, would be undignified vis-à-vis Kant. It would also be pointless with regard to the goal we are striving at. We want to attain a positive clarification of the concepts of Dasein, existence, and being as such, and we want to attain it such that we do not just posit our own opinion contrary to Kant's, as an external position. Instead we wish to develop Kant's approach further, the interpretation of being and existence, and to do so according to his way of looking at things. Ultimately, Kant is surely heading in the right direction in trying to elucidate Dasein and existence. Only he fails to see the horizon with sufficient clarity, from where and within which he intends to conduct the clarification. He does not see it because he does not survey this horizon properly and because he does not explicitly place his elucidation within this horizon. (*Gesamtausgabe* 24:64; my translation)

While the scholarly path is marked by observation, preservation, and obedience, in a word, by *gentleness*, this trail is rougher. Scholarly inquiry is minimally invasive; the synthetic inquiry is more probing, more proactive. The initial criticism—a resounding no—reflects the intensity of this approach. The synthetic path, then, does not end on a critical note; it *begins* with it. Its first step is an antithetical positioning, a considered negation of the text. This gesture of dissent frees the reader from reverence, but only to serve a heuristic purpose in line with the original thrust of the text.

Thus conceptual synthesis begins with a gesture of dissent made in the context of affirmation. The antithetical positioning is a denial of concrete textual assertions. The context of affirmation is the recognition of the Kantian perspective in general, not as an ideological stance, but as an acknowledgment of Kant's status in civilization. Many thinkers try to make contributions; Kant is one of the few who did. He belongs to a select group of perennial thinkers on par with the set of scientific pioneers. Pioneering contributions, by definition, cannot be perfect, and critique and revision are needed for making progress. Perennial contributions are not perfect either. Their perennial character consists in recurrent fertility, in disclosing fresh meaning, time and again in different epochs and to subsequent generations. This is the context of affirmation. Conceptual synthesis is accordingly a middle way between dogmatic and skeptic extremes, between the extremes of elevating a text to holy writ and dismissing it out of hand.

Such a balance between dogma and doubt allows for an inquiry that seeks to unearth the deep substance of the text. A skeptic would scoff at this hope as being naïve, deeming it romantic at best and deluded at worst. A scholar would reject it as being incoherent; from a scholarly point of view, the substance of the text is its literal content; you see what you get, and there is nothing 'deeper' than that. A Kantian might not feel comfortable with this aspiration either, since it tends to leave literal content behind for the sake of risky interpretive appropriations. Conceptual synthesis would displease skeptics, scholars, and Kantians alike. The synthetic way of engaging with past masters is like a hermeneutical dance with a classic for the sake of disinterring its living information; a summoning of old ideas to conjure up new insight. In a manner of speaking, conceptual synthesis is rational necromancy or *Geisterbeschwörung:* past master and present reader enter in a virtual discussion.

Different from Heidegger is that the reader does not act as a mere interrogator in this necromantic dance. The reader goes to the text not just with questions. Scientific progressions since the time the text was written guide the reading of classical claims. In conceptual synthesis readers approach the text armed with facts. Progress affects philosophy retroactively, for scientific findings settle age-old metaphysical disputes. Some questions do have answers, even those considered as perennial problems. In this dance of ideas and facts around core issues, the past master is not always the leader; sometimes readers take over.

The solution to the riddle of the power of imagination in moral cognition now comes into view. That Kant denies the use of the imagination in the moral litmus test requires a closer and critical look—not for making a different point, but for updating Kant's contribution in light of present information. Common sense should make us suspicious about Kant's denial. The sequence of procedures for testing the maxim of false promises reveals that imagination or its identical twin is at work here. The third step of the test is crucial. Recall this step as the universal actualization of the potential, the radiation of seen individual results across the collective, and the vista of consequences all the way to the end. Suppose everyone acted 'that way,' what would happen to the collective? The answer (there 'would properly be no promises at all') hinges on envisioning the supposition, extending this vision across space (vis-à-vis everyone), and chasing this spatial vision through time (toward consequences). Calling this cognitive performance a work of the understanding is not convincing. The role of the understanding according to the first Critique is that it produces concepts—not visions. This is even clearer in the shift from step two to step three of the litmus test. Completing the transition from envisioning *oneself* as making a false promise, as suffering lost credibility, and as risking revenge, to envisioning *others* as doing the same requires the multiplication of the individual vision to a collective vision. One image—seeing oneself as making a false promise—is first multiplied into many and next linked together as a manifold vision: seeing *multiple selves* making false promises, and seeing them *all together.* The completion of this transition shows two traits of the power of imagination at work, painting images and forging links. Akin to the faculty of intuition, the power of imagination produces images, visions, or scenarios. And similar to the faculty of reason, imagination joins, connects, and adds things together to a larger interactive whole. That's the imagination. This is what it does in the moral litmus test. If it walks like a duck, and if it quacks like a duck, well then, call it a duck. Calling the imagination by another name, 'understanding,' arguing that it is not involved when in fact it is, or labeling its product as 'type' instead of 'schema,' is simply misleading. It's a duck.

4 Progress

The motivation for using the understanding in the moral litmus test is the transcendental ideal of the good. The good is said to be necessary and universal; it is held to be formal and part of the intelligible world. So the litmus test depends on the cognitive faculty that deals with intelligible entities such as concepts, the understanding. The problem with inserting the understanding is, of course, that there is no evidence for this claimed loftiness of the good. The supernatural notion of the good is just a hope.

There is nothing wrong with hope, but scientific progress since Kant's time indicates that the good is more comfortably situated in the natural world than Plato and monotheists would allow. The preconditions of moral agency, rationality and free will, have been demystified: as D. Dennett describes, today we know that freedom and judgment evolve. With the evolution of these preconditions, moral agency comes into being as the product of self-organization. Since the work in sociobiology by E. O. Wilson we know how moral behavior arises in complex groups. We must recall that culture itself is an expression of nature. The primatologist K. Imanishi has found cultural variability in nonhuman communities such as those of macaques; these old world monkeys exhibit regional differences over intergenerational learning. The primatologist F. de Waal has found that justice, the essence of moral agency, is a concept we share with macaques as well. Other work shows moral agency in other species. The good is perfectly natural.[2]

---

2 There is much literature on the evolutionary roots of morality, the natural grounding of the good, as well as on nonhuman rationality, fairness, and culture. Key publications in the history of science are, re culture, Frisch, K. v. (1923): "Über die 'Sprache' der Bienen: eine tierpsychologische Untersuchung". In: *Zoologische Jahrbücher* 40 : 1–186; and the report by a member of Imanishi's research team, Kawai, M. (1965):"Newly acquired pre-cultural behavior of the natural troop of Japanese monkeys on Koshima islet". In: *Primates* 6: 1–30; re nonhuman thought

The confusion about the terms 'natural' and 'supernatural' adds to the riddle. What is contingent and particular is the opposite of what is necessary and universal. On an epistemological level, we associate the former with sensible, material structures, what Kant calls 'nature,' and the latter with intelligible, formal structures, what Kant ascribes to the noumenon, and cognize the two structures as being mutually exclusive. This distinction informs the epistemological dualism between the sensible and the intelligible. But just because we distinguish concrete facts from universal truths in our cognition does not entail that they inhabit different worlds. What we can know one thing, reality is another. Epistemological dualism, however warranted, does not entail ontological dualism.

As soon as one keeps ontological and epistemological dualism apart, the ambiguities in the moral litmus test about the cognition of the good—as image and as concept—clear up. On one level (the level of hope), the good emerges as an intelligible absolute, a transcendental ideal reflected by reason and conceptualized by the understanding. On another level (that of collective behavior), the good emerges as naturally sustainable functioning, which is accessible to imagination. Both levels belong to the litmus test. The supernatural idea of the good is the horizon and frame of the test. The natural image of the good as sustainable functioning occupies the center stage and concerns the core of the test. So we can grant to Kant that the power of understanding is involved. But the involvement of the understanding does not compel us to throw out the power of the imagination. There is no conflict. Both have work to do; there is only a division of labor. Understanding (as Kant writes in the *Critique of Practical Reason*, 5:69) can put the formal reflection of the moral law under the idea of reason. In the test, this is the vanishing point at the horizon. Imagination (as can be added—conceptually synthesized—to *Groundwork* 4:403) can inform us in the shortest and yet infallible way about which envisioned scenarios, which simulated spacetime floats, are moving in straight lines to this horizon point. Understanding allows us to perform the test in principle; it makes possible the shell of the test: step one and four, the query and the judgment. Imagination allows us

---

contents: Carmena, J. M. et al. (2003):, "Learning to control a brain-machine interface for reaching and grasping by primates". In: *PLOS Biology* 1 : 193–203; re nonhumans and the negative golden rule: Brosnan, S. and F. de Waal(2003): "Monkeys reject unequal pay". In: *Nature* 425 : 297–299. For a non-technical overview of the evolution of the preconditions of moral agency cf.. Dennett, D. C. (2003): *Freedom Evolves*. New York: Viking. For an introduction to sociobiology cf. Wilson, E. O. (1980): *Sociobiology: the New Synthesis*. abridged ed. Cambridge, Mass.: Harvard University Press.

to perform the test in practice; it makes possible the core of the test: step two and three, the personal scenario and the collective scenario.

Rehabilitating the imagination in moral cognition is a rather technical concern, but it sheds light on larger questions. Progress, in both practical and theoretical meanings, is a leitmotif in Kant. How citizens can make progress toward an enlightened age is the theme of *What is Enlightenment*. What progress means in terms of global politics and international law is spelled out in the articles of *Perpetual Peace*. Why metaphysics has made no progress and how it could enter 'the secure course of a science' is a key issue in the *Critique of Pure Reason* (B vii, tr. Guyer). The introduction culminates in five questions (B19-B22), about the possibility of the synthetic a priori, of mathematics and science, and of metaphysics. What Kant does not question, but what we should not take for granted, is the leitmotif itself. Metaphysics, science, and cognition are being queried on their potential to proceed to knowledge. The actuality of progress is assumed, but nowhere is its *possibility* examined. So: how is progress possible?

Specifically, how is progress possible in practical terms, in the moral realm? Its meaning is essentially a progression of civilization along what Aldo Leopold called *the ethical sequence*.[3] This progression manifests itself in social, political, and cultural advances. Examples are the change of the legal status of normal people from serfs to citizens during Renaissance and Enlightenment; the church-state division and the division of state power into branches during the Enlightenment and the nineteenth century; the abolition of slavery and the creation of the International Red Cross in the nineteenth century. The twentieth century, despite all its mayhem, enjoyed a breathtaking accelerando along the sequence—from the Geneva Convention at one end to the Kyoto Protocol at another, from soldiers' rights to women's rights to racial equality to decolonialisation to animal rights to a regulation of the commons. The ethical sequence marks progress toward the tripartite ideal of *egalité, liberté, fraternité*. It also marks progress on a larger scale, the move from a savage, exploitative starting point to a sustainable goal state. For Aldo Leopold, this is the change

> [of] the role of *homo sapiens* from conqueror of the land-community to plain member and citizen of it. It implies respect for his fellow-members, and also respect for the community as such. (Loc. cit., 204)

Thus progress occurs. Today, this envisioned change of the human role has become a necessity; progress, now, means the transformation of climate-changing

---

3 "The Land Ethic," orig. 1949, cf. esp. p. 202–203 in: Leopold, A. (1989) :*A Sand County Almanac*. Oxford/New York: Oxford University Press.

consumer societies to a sustainable world. In the future, progress is bound to continue on Leopold's path, since the crossing of sustainable yield thresholds and the overwhelming of system capacities push the planet to the brink and force the hand of civilization. Progress towards sustainability is the only way global civilization can prevail.

Empirically, progress happens in fits and starts, leaps and bounds. Along the ethical sequence, progress is not smooth. Each leap is triggered by a swelling of discontent towards a critical mass. Its explosion unleashes revolutionary energy that leads to what Alain Badiou calls *events*—the fateful moments in history when everything hangs in the balance, everything is possible, anything can happen. And indeed: anything *can* happen. Occurrences that are unimaginable in advance snap into reality in such a way that they become the new normal and are taken for granted—a miracle morphs into routine.

The German reunification in 1990 was the outcome of a revolutionary uprising in East Germany. The uprising took off from weekly peace prayers held in Leipzig's Nikolai Church. The Monday prayers began in 1982; in 1989, over a few weeks in mid-autumn, these vigils spilled out of the church doors over into the square and down the streets. They became known as the *Montagsdemonstrationen*, the Monday demonstrations. Arrests began on September 11, when the mass had grown to a thousand participants. On the following Mondays, arrests multiplied, but the crowd grew even faster. Twenty thousand gathered on October 2, and the police reacted violently. On October 9, two days after the 40th anniversary of the founding of the German Democratic Republic, a crisis was in the making. City hospitals stocked up on blood reserves; eight thousand military troops and police officers descended on Leipzig, where seventy thousand protesters assembled. In the morning, the government debated regaining control by resorting to the *Chinesische Lösung*, the 'Chinese Solution' (the Tiananmen Massacre June 4, 1989). At noon, city leaders in Leipzig appealed to common sense, courtesy and nonviolence—and joined the demonstrators. In the afternoon, the protesters began to march. Then the event happened: when the people reached union station at 8:00 pm, the police began to retreat: the power of the regime was broken. On October 16, one hundred thousand met; on October 23, two hundred thousand; on October 30, three hundred thousand, and on Monday, November 6, half a million were in the streets. Three days later, the Wall came down.

Any revolution is a leap of unlikely proportions. What makes the 1989 East German revolution so stunning is that it was the overthrow of an armed dictatorship without incurring a single fatality. It was the first pacifist triumph over tyranny in the history of civilization.

And yet it happened. It happened because ever larger numbers of citizens *imagined* it. The dispiriting power of a totalitarian regime was met with increasing resentment, but out of this bitterness came hope, and this hope was guided by a vision—a forecasting of a possible scenario in the mind. The phenomenology of imagining as progressive forecasting is perfectly captured in the opening track of John Lennon's 1971 album *Imagine*. The lyrics are mundane yet precise. Lennon sings:

> Imagine there's no heaven / It's easy if you try / No hell below us / Above us only sky / Imagine all the people living for today / Imagine there's no countries / It isn't hard to do / Nothing to kill or die for / And no religion too / Imagine all the people living life in peace / ... / Imagine no possessions / I wonder if you can / No need for greed or hunger / A brotherhood of man / Imagine all the people sharing all the world / You, you may say / I'm a dreamer, but I'm not the only one / I hope someday you'll join us / And the world will live as one.

It is useful to complement this poetic profile of the imagination as a key condition of progress with a remark Kant makes in *Perpetual Peace*. Imagining is not the exclusive domain of hippies; neither is it the prerogative of ordinary people who are simply disappointed in dysfunctional leadership. It serves as the foundation of moral cognition and as driver of Leopold's ethical sequence precisely because it is not itself a moral aspiration at all. Political imagination is meaningful even to a 'nation of devils'. All that is needed is the understanding, in addition to imagination. Kant writes:

> The problem of establishing a state, no matter how hard it may sound, is *soluble* even for a nation of devils (if only they have understanding) (*ein Volk von Teufeln (wenn sie nur Verstand haben)*) and goes like this: 'Given a multitude of rational beings all of whom need universal laws for their preservation but each of whom is inclined covertly to exempt himself from them, so to order this multitude and establish their constitution that, although in their private disposition they strive against one another, these yet so check one another that in their public conduct the result is the same as if they had no such evil disposition.' Such a problem must be soluble. For the problem is not the moral improvement of human beings but only the mechanism of nature, and what the task requires one to know is how this can be put to use in human beings in order so to arrange the conflict of their unpeaceable dispositions within a people that they themselves have to constrain one another to submit to coercive law and so bring about a condition of peace in which laws have force. (8:366.15 – 29; tr. M. J. Gregor 1996: 335)

Understanding (*Verstand*) is the organ of rationality that allows us to find a technical solution to the problem faced. The problem is that self-preservation requires an order that conflicts with inclination. The solution is to design a legal structure that funnels individual inclination into collective welfare. Just as in the moral litmus test, understanding contributes the frame—the question or problem and the answer or solution. Not mentioned here, but necessary as well, is the imagination that allows us to cast an image ahead through space-time, a vision that serves as a rational goal-state. This vision, for rational devils, citizens, and hippies alike, is the futurist telos of lawful communities in peaceful arrangements. On the moral level, the generation of visions drives the internal machinery of the litmus test; on the political level, it is the incentive of progress, both as the sheer energy of activism and as the destination of political change.

The discovery of the power of imagination as the driver of the moral litmus test thus gives us the key to recognize the cognitive dimension of progress, and hence its possibility, on the sociopolitical level. Over practical matters, whether in morality or politics, understanding and imagination work together. Their co-operation is the interplay of a faculty and a power. The faculty of understanding imposes logical structure on normative decisions and political action. The power of the imagination evokes scenarios and conjures visions, injecting a dynamic element. Moral decision-making and sociopolitical progress are thus structural-dynamic phenomena. Kant's attention is directed to structure, and so it is no surprise that the understanding swells to unrealistic significance, obscuring the imagination in his practical philosophy. No wonder: the builder of the critical architectonic sees only brickwork, but not the mortar within.

5 Evolution

Let us recap. The power of the imagination, foundational and elusive in the critical system, cannot be clarified by scholarship alone. If we want to determine the role, nature, and roots of this power, empirical analysis must yield to conceptual synthesis. The mismatched pieces of textual information pose riddles, such as the ethical emphasis on the understanding at the expense of the imagination, which flies in the face of contrary evidence, and the uncritical notion of the possibility of progress, which does not square with its function as a critical leitmotif. Solving these riddles will require realigning the pieces on an interpretive canvas stretched over four sides: the gesture of dissent, the context of affirmation, the fertility of perennial insight, and the findings of science. This canvas lets us re-insert the imagination into ethics. This projection cannot fill the gap between the first and the third Critiques because Kant explicitly expels the imagination from the second Critique, but it puts his mistaken expulsion in perspective—when seeing the text on the canvas, analytic data versus conceptual synthesis, we gain stereoscopic perception. This outcome is a not so much depth perception in space but rather a parallax view in time: from the historical artifact of Kant's fractured but seminal philosophy to a future-oriented project of appropriation.

The progressive-political role of the imagination shows it shuttling to and fro between intuition (*Anschauung*) and intellect (both understanding and reason). It steals images from the receptivity of impressions to jam them to the spontaneity of concepts, smashing up imaginative ideas of varying levels of complexity, pumping out goofy, mutable, but sometimes brilliant scenarios. Evoking these visions is a dialectic activity. It is analysis as well as synthesis, pulling images from sensibility, concepts from understanding, and ideas from reason, and putting them together into a free-spirited, meaningful whole. It involves affirmation, by absorbing impressions that, on the sociopolitical level, can be traumatic, and negation, by rejecting them the imagination and defiantly flipping them upside down. Faced with a wall, we imagine there is none and tear it down. Faced with war, we imagine no one attends and keep the peace. Faced with consumerism, we imagine communist modesty and create a sustainable future.

Imaginative dialectic is analytic yet synthetic, affirmative yet negative, and always progressive. The imagination-shuttle zips about. In time, the power of imagination casts a visionary line ahead of itself, flings it to tomorrow, loops it around a peg (a beacon of hope, a stake of horror), and hauls the line back to the present, so as to permit progress to winch today toward the future. In space, this power braids scenarios by pulling out pictures from here and pushing them over there. The action in time transfers an imaginative impulse from the moment to some possibility, unfurling a visionary tapestry that travels until it lands in some tomorrow, from where it is reeled back from the future, down to the present. The imagination casts and hauls. The spatial action weaves from here to there and back again. It tears out impressions from the senses and glues them together into fantasies. Whether as casting and hauling, or as tearing and gluing, the kinetics is clear. Utterly naked, imagination is a pulse of push and pull.

This pulse is not only the dynamic essence of the imagination at work; it is also the evolutionary engine in Kant's early philosophy of nature. In his first book, *Living Forces*, it pumps out spacetime and matter (cf. 1:18–24; § 1–10). In his second book, *Universal Natural History*, it pumps out the seesaw of coastal winds, solar systems, galaxies, and the cosmos (cf. 1:223–234, 234, 264–269, 1:306–322; preface, chapters 1 and 7). In his professorial thesis, *Physical Monadology*, it pumps out the force fields whose event horizons are constitutive of spatial and material extension (cf. 1:481–482, prop. 7). Some of these pre-critical conjectures anticipated contemporary theorizing in quantum cosmology and string theory, others have found confirmation by later discoveries in climatology, stellar dynamics, and astrophysics.[4]

Kant calls evolution *die Auswickelung der Natur* (1:226.8), 'the out-wrapping of nature,' a literal rendition of the Latin *evolvere*, which means to unroll, and whose noun *evolutio, -onis* means reading books, literally the unrolling of a parchment. It suggests an idea of evolution both older and younger than Darwin's; evolution as self-organization, unfolding from chaos to order, from entropy to complexity.

So the natural power of evolution and the cognitive power of imagination share a dynamic pulse. The power of evolution works within nature to unfold nature. The power of imagination works within cognition to unfold cognition. It does so for particular cognition, as in deciding what maxims are right and in envisioning goals for progress, and for general cognition, by unifying the sensible consciousness, described in the *Critique of Pure Reason* as "that which connects the manifold of sensible intuition" (B164). The imagination-pulse puts apprehension together, effectively making consciousness possible.

For empirical-analytic scholars, this correlation of natural and cognitive pulse is inconsequential. Since Kant does not explicitly identify the two pulses in the known oeuvre, further study is discouraged. Probing deeper would violate the rule of observation. In the absence of textual data, scholarly constraints terminate inquiry. For the analytic mind, the dual pulse is a coincidence, nothing to lose sleep over.

Not so in conceptual synthesis: here the correlation cries out for interpretation. Is there a link? The question is of an ontological kind: does the twin pulse disclose a path from nature to consciousness, from the power of evolution to the power of imagination? Or, to reshuffle the question: is there an evolutionary path from nature to consciousness *that leads through the imagination?*

To see how the imagination may have been involved in the leap towards consciousness, recall what it does in such mundane activities as daydreaming. It uses residues from impressions and distillates from experience, and splices

---

4 For more on these claims, cf. Martin Schönfeld, "Kant's philosophical development," *Stanford Encyclopedia of Philosophy*, 2003, online; "Kant's thing in itself or the Tao of Königsberg," *Florida Philosophical Review* 3 (2004): 5–32; "Superstrings and the Euler-Kant mirror," *NCCU Philosophy Journal* 13 (2005): 99–124; "Kant's early dynamics," 40–60 in G. Bird, ed., *A Companion to Kant* (Oxford: Blackwell, 2006); "Kant's early cosmology," 61–84 in G. Bird, ed., *A Companion to Kant* (Oxford: Blackwell, 2006); "The phoenix of nature—Kant and the big bounce," *Collapse* 5 (2009), 16 pp., and "Factual Notes to *Thoughts on the True Estimation of Living Forces*," in Kant, *Scientific Writings*, vol. ed. E. Watkins, in ed. P. Guyer, A. Wood, *The Cambridge Edition of the Works of Immanuel Kant* (Cambridge: Cambridge University Press, 2012, in press)

them together in a sequence that aspires to be realistic as possible while roaming free. Is daydreaming of evolutionary consequence? It can be, if imagination works in the way of the moral litmus test, with the goal of seeing in advance likely causal cascades triggered by a possible course of action; if it determines, in short, the consequences of a deed without getting one's hands dirty. In daydreams of evolutionary consequence the imagination works by running *simulations*.

Simulations, in evolutionary terms, are a matter of life and death. Existentially, simulations serve the end of saving one's own hide and to avoid winning a Darwin Award. Running them lets an embodied mind avoid epic failure. The biologist Richard Dawkins notes,

> Survival machines [e.g. organisms] that can simulate the future are one jump ahead of survival machines who can only learn on the basis of trial and error. The trouble with overt trial is that it takes time and energy. The trouble with overt error is that it is often fatal. Simulation is both faster and safer.[5]

Think, then, of a thirsty deer scenting water. Or think of two deer. The one follows the scent to the water hole and starts drinking. The other hesitates. Water holes draw also predators. Perhaps the reluctant deer has seen a kin being eaten there. How big is the cognitive step of blurring the traumatic memory of the predator attack into a vision of falling prey? Look again: there she runs, bends down to drink, touches the surface, jaws snap and pull her in; she screams, thrashes, and is gone. The step from a terrible memory to a terrifying vision involves but the smallest of twists. It is the insertion of a simple imaginative fear: *this could be me...*

Closing the connection between oneself and another, between witness and victim, happens *in* the mind, but there is no need to assume it must happen *by* the mind. Making the mind responsible, as an empowered, deliberate agent, turns the connection into an unnecessary riddle. How can a dumb deer wise up? How can herbivores *make* an imaginative leap? If we took the connection as a performance done spontaneously, freely, and inferentially by a ruminant mammal, we would attribute a quality to the link that is different from more basic cognitions in the mental flow. But there is no need to impute so much. The event of a link being made could be of the humble sort shared by human and nonhuman cognitions alike: a not-quite conscious weaving of data into emergent information, a creeping coalescence of images and feelings into

---

5 Dawkins, Richard (1989): *The Selfish Gene*. Oxford/New York: Oxford University Press, orig. 1976, 59.

an insight—the way in which comprehension often dawns on us. This is simpler, and may be more appropriate. The connection *is being made*; and there is no agent making it. It just happens.

Identifying one's being with that of another can thus emerge as an incremental or *dawning* realization. In the end the realization is completed; at that instant one does identify with another, and from this point onward one *can* do this. The cumulative realization itself hinges on a data invasion from the environment to the mind. This is why identification with another appears as an action, while it is originally driven by a passive cognitive flow. This flow brings about the creation of a capacity, whose subsequent exertion is indeed activity, but the creation itself turns on the reception of impressions.

The trauma of witnessing another being seized, dragged, and eaten is not just the sight of it. The impressions are also auditory, with screams, grunts, splashes, and yelps, and may also assail the nose. Olfactory data can linger as the metallic scent of blood and the acrid stench of fear. A scientific finding after Dawkin's conjecture concerns the way smells cause identification in sheer physiological terms: the nasal reception of chemosensory anxiety signals triggers a physiological adjustment in the brain *without being dependent on conscious mediation*.[6] The adjustment mirrors the condition witnessed; the witness grows anxious and emits the same chemosensory signals. Thus fear spreads, and panic can seize a herd.

The chemical bridge from victim to witness is the linchpin. Here is another being, evidently in extremis, in obvious distress. The sensory assault by this datum affects the observer with empathic distress. Now the sensory reception of the discomfort witnessed invades the observing subject on a holistic level: it is not an isolated visual or auditory datum anymore, perceived at someone else, *but now transforms into a systemic physiological response*. The experience is now suffered in the observer. In perceiving the distress, I am becoming distressed as well; in witnessing the plight of the victim, I am appropriating the plight and *make it my own*. Now my heart beats fast too; my limbs go numb; I want to run away. A connection is made: I am the other.

Dawkins (loc. cit.) continues:

> The evolution of the capacity to simulate seems to have culminated in subjective consciousness. Why this should have happened is ... the most profound mystery facing biology. There is no reason to suppose that electronic computers are conscious when they simulate, although we have to admit that in the future they may become so. Perhaps consciousness

---

6 Prehn-Kristensen, A. et al.(2009): "Induction of Empathy by the Smell of Anxiety". *PLoS ONE* 4 , e 5987, p. 8.

arises when the brain's simulation of the world becomes so complete that it must include a model of itself.

Imagination saves lives. Deer that can make the link have better odds at reaching reproductive age and handing down this peculiar sensitivity to their offspring. Dawkins (loc. cit.) concludes:

> Whatever the philosophical problems raised by consciousness, for the purpose of this story it can be thought of as the culmination of an evolutionary trend towards the emancipation of survival machines as executive decision-takers from their ultimate masters, the genes.

In lending a competitive advantage, imagination may be the root of consciousness. Through daydreaming activities such as running simulations, the imagination is a bootstrap device. When life evolves so as to individuate into experiencing centers, an organic level of sentience is reached. At this stage there is a sensory center, a proto-consciousness, but no sense of self. A further step leads from sentience to consciousness, from sensory centeredness to personal identity. This step is made by the imagination. A short-circuit of simulation and empathy lets one feel with others and thereby gain a self.

In sum, Kant's philosophy sheds light on the deep structure of the imagination—its roles, nature, and roots—but only if one approaches the Kantian use of the imagination in a constructive, synthetic, and creative manner. More than any other concept, the imagination in Kant points to the limits of analytic scholarship. We went beyond these limits by connecting Kant's critical view of the imagination with his pre-critical view of evolution, and by rehabilitating the role of the imagination in the social sphere, which includes ethics and politics and generally concerns the functioning of communities. Both such a connection and this specific rehabilitation are in direct violation of Kant's own word. But they remain arguably in the spirit of a pure Kantian perspective. Aided by such connection and rehabilitation, it is possible to construct a unified account of the power of imagination. This unified account integrates general cognition, moral cognition and heuristic progressions. The heuristic progressions made possible by the imagination, in turn, point to the viability of evolutionary leaps in the animal kingdom. Dawkins suspects that the evolution of this capacity for simulations culminates in subjective consciousness. This is suggestive but leaves a gap: how, specifically, can one get from the one to the other? How does running simulations in general get us to the acquisition of a self? Dawkins suggests one intermediate step: the acquisition of the self is the outcome of running simulations of one's environment with one's own person in it. This suggestion raises immediately another question: how does running simulations in general lead to simulations with one's own person in it? I suspect that imagination is the bridge to personhood: a mirror-like identification with the plight of others through a chemically triggered physiological response. If this is true, then the power of imagination will indeed be the evolutionary root of consciousness. And if this is the case, then it will have become clearer why the power of imagination must play such a strange role in the geography of the mind of the *Critique of Pure Reason:* pervasive and yet elusive, vital and yet subterranean, central and yet obscure. If consciousness is the evolutionary flower that blossoms from the stem of the imagination, then it hardly could be otherwise.

Rudolf Makkreel

Recontextualizing Kant's Theory of Imagination

Ever since Immanuel Kant, the imagination has come to be recognized as not simply an arbitrary source of fancy, but as a power that can be productive in the acquisition of knowledge and creative in the appreciation of aesthetic order. He assigned the imagination an important place in his philosophy and over the years its functions evolved. In his pre-critical writings, Kant considered imaginative formation (*Einbildung*) in relation to a range of other empirical formative functions. Then in the *Critique of Pure Reason*, he gives the imagination (*Einbildungskraft*) a central synthesizing role in the production of knowledge. Although most commentators extend these synthesizing functions to the aesthetic imagination of the *Critique of Judgment* as well, I have argued that the transition from determinant judgment to reflective judgment demands other functions from the imagination as well.[1] Here this thesis will be extended to propose that we can reassess the function of the imagination in relation to the different contexts in which it comes into play. I will point to Kant's views on orientation as providing a basis for this adaptability thesis. By also taking note of Wilhelm Dilthey's innovative theory of mind, it can be made clear how the imagination can play a hermeneutical role.

Starting with Kant's views on image formation as needed for the perception of things and on imaginative schematization as needed for the apperception of objective experience, I will develop other uses of the imagination that prove to be useful for interpreting the world. They explore the way we relate to the world as our overall frame of reference. To the extent that we are already generally oriented to the world at large, the imagination can highlight more specific modes of connectedness. But when the demands of life lead us to confine our attention to local contexts, the imagination becomes essential as the capacity to configure other possible contexts. The hermeneutic function of the imagination is to further orient us by recontextualizing our experience.

---

1 See Makkreel Rudolf A. !990):, *Imagination and Interpretation in Kant*; *The Hermeneutical Import of the* Critique of Judgment. Chicago: The University of Chicago Press, ch 3. See also "Response to Guenter Zoeller's review-essay, 'Makkreel on Imagination and Interpretation in Kant'". In: *Philosophy Today*, vol. 36:6 Fall 1992, pp. 276–280.

Image Formation and Spatial Gathering

In his *Lectures on Metaphysics* from 1778 – 80, Kant focuses on imaginative formation (*Einbildung*) by comparing it to a range of empirical formative functions such as direct image formation (*Abbildung*), reproductive image formation (*Nachbildung*) and anticipatory image formation (*Vorbildung*). These three empirical modes of image formation are considered as products of *Imagination,* which is the "storehouse (*Vorrath*) of representations."[2] Even anticipatory image formation is conceived as a product of empirical laws of association: it merely extrapolates from the past and present to the future. Only with *Einbildung* does the spontaneity of consciousness begin to exhibit itself. It is defined as "the capacity to produce images from out of oneself (*aus sich selbst Bilder hervor zu bringen*) independently from the actuality of objects."[3] *Einbildung* is thus more akin to completing formation (*Ausbildung*) and counter-reformation (*Gegenbildung*), which also depart from what is given to us in intuition. *Ausbildung* projects "an idea of the whole"[4] to complete what is left incomplete in experience. *Gegenbildung* is defined as the capacity to characterize in the way that words can represent things in their absence. *Gegenbildung* will later be related to how Kant conceives of symbolization in the *Critique of the Power of Judgment.* There we can speak of symbolic analogues that produce indirect links between different spheres such as sense and reason which cannot be directly linked.

Although *Abbildung* is said to be dependent on how our senses are affected by objects, Kant nevertheless conceives it as actively formative. He writes in the *Lectures on Metaphysics:*

> The mind must undertake to make many observations to form a direct image of an object. This is because it forms a different image from every side... There are many different appearances of a thing from different sides and viewpoints. From all these appearances, the mind must make itself an image by gathering them together.[5]

What Kant expects from *Abbildung* is a synoptic image that gathers different perspectives together to represent the spatial presence of an object. This *spatial synopsis* of sense seems to be the point of reference when Kant goes on in the Sub-

---

2 Immanuel Kant, *gesammelte Schriften, herausgegeben von der Preussischen Akademie der Wissenschaften zu Berlin* (hereafter Ak), 29 vols., (Berlin: Walter de Gruyter, 1902 – 97), 15: 132
3 Kant, Ak 28:237
4 Kant, Ak 28:237
5 Kant, Ak 28:236

jective Deduction of the *Critique of Pure Reason* to speak of a corresponding need for a *temporal synthesis.*[6]

In the *Critique of Pure Reason* of 1781 Kant calls the synopsis of sense that "contains a manifold of intuition" receptive. The formative activity of spatial gathering ascribed to *Abbildung* as late as 1780 is now denied the spontaneity needed for cognition. The gathering function of *sensible Abbildung* is assumed to occur unconsciously and to have its conscious counterpart in a synthesis of *intuitive* apprehension that runs through the manifold in a temporal succession. This is how Kant puts it: "Every intuition contains a manifold in itself, which however would not be represented as such if the mind did not distinguish the time in the succession of impressions on one another."[7] An *a priori* synthesis is needed to become conscious of the manifold *as* a manifold while also apprehending it as a unity.

Imaginative Synthesis and Temporal Recognition

The unity of the synthesis of apprehension is not yet a cognitive unity but preliminary and intuitive. The Subjective Deduction adds two more syntheses that involve imaginative reproduction and conceptual recognition respectively. We find that the imagination contributes directly to the reproduction of representations and indirectly to the productive syntheses of representations needed to attain adequate knowledge of objects. According to Kant, mental representations are ephemeral and must therefore be actively reproduced from one moment to the next. When representing a complex object we must reproduce those parts initially apprehended while advancing to others that follow if a complete representation is to be obtained. The reproduction of partial representations over time is a necessary condition for eventually reaching an adequate representation of a complex object, but it is not a sufficient condition. The imaginative synthesis of reproduction keeps alive the initial representation while at the same time associating it with succeeding representations. But imaginative reproduction would be in vain if we could not recognize what is reproduced at time 2 as a reproduction of what was first apprehended at time 1. This act of recognition is a productive synthesis that identifies the sameness of representation 1 over time

---

6 For a more detailed explication of these early discussions of image formation, see Makkreel, *Imagination and Interpretation in Kant*, ch.1.

7 Kant, (1997): *Critique of Pure Reason*, ed. and trans. Paul Guyer and Allen Wood, Cambridge UP, A99

while linking it with representations 2, 3, etc. that also refer to the object being perceived.

The productive synthesis that makes possible the transition from mere subjectively associated representations to objectively unified representations requires a concept according to Kant. As the Objective Deduction of the first edition already makes clear, the imagination can only be cognitively productive in conjunction with the understanding. Nevertheless, the imagination is indispensible for cognition because it alone can form the schemata that mediate between universal concepts about objects and the particulars given by sense. What Kant has done is to show that the imagination plays a central cognitive role, not by generating spatial images of objects, but by forming temporal schemata for them. A schema applies a universal concept of the understanding to the most pervasive condition of sense, namely, the form of time. It translates the atemporal rules implicit in a concept into a temporally ordered set of directions for referring a sequence of mental representations to objects of sense. Cognitively, the understanding *seeks* and the imagination *finds*. That is, the understanding *seeks* objects for its concepts, but we need the imagination to give us the directions for *finding* them. Without the aid of the imagination the pure concepts of the understanding would remain intuitively empty according to Kant and lack the meaning (*Bedeutung*) that allows them to point or refer to (*deuten*) specific objects. What he calls the "figurative imagination" in the first *Critique* really *prefigures* the referential meaning that objects can have for us. Thus the imagination provides the directions that can give the pure concepts of substance and causality their objective meaning. It does so by modulating from the medium of thought to the medium of sense.

This account of the productive and schematizing functions of imaginative synthesis has proved to be so powerful that the task of synthesis is commonly transferred to the aesthetic imagination of the third *Critique* as well. But Kant's aesthetic imagination is no longer the handmaiden of the understanding and "schematizes without a concept."[8] Since synthesis requires concepts, I have argued that a non-conceptual aesthetic imagination cannot synthesize.[9] But if the aesthetic imagination does not produce a synthetic objective meaning, does it fall back into mere subjective associations? For the freedom of the aesthetic imagination not to collapse into arbitrary associative fantasy, it must be capable of producing at least an intersubjective meaningfulness. I propose

---

8 Kant, Immanuel (2000): *Critique of the Power of Judgment*. Cambridge: Cambridge University Press, §35; Ak 20: 234.
9 Makkreel, *Imagination and Interpretation in Kant,* chapter 3

that if schematizing with a concept allows the imagination to direct us to objects and *give* them meaning, schematizing without a concept allows it to *share* meaning in a more general way. In §9 Kant contrasts a free aesthetic play of the imagination and understanding, qua faculties, with the "objective schematism" of the cognitive judgment in which the understanding guides the imagination with a concept. Without concepts we cannot have discursive communication about the contents of sense, but Kant assigns the free non-conceptual play of the imagination and understanding a "universal communicability"[10] that applies to our felt states of mind. This clearly indicates that the schematizing without concepts referred to in §35 involves an aesthetic communicability whereby states of mind can be shared. The communicative function of the aesthetic imagination does not contribute conceptually to our experience of particular objects, but allows us to feel attuned with the world of human subjects.

Instead of speaking of imaginative synthesis in the *Critique of the Power of Judgment*, Kant points to the aesthetic imagination's harmonizing effect on feeling. Here the schematizing power of the imagination harmonizes our experience. It does not produce objective cognition but an intersubjectively valid evaluative attitude that finds its confirmation in a formal feeling of pleasure derived from a sharable harmony of the human faculties. I will have more to say about how pure schematization can harmonize our experience in relation to the idea of an orientational imagination.

Both the epistemic and aesthetic functions of the imagination in Kant serve to relate mental representations that are assumed to be initially discrete. This atomistic assumption is questioned by Dilthey who conceives consciousness as a continuum or living nexus. Even our first impressions of the world around us have a connectedness. The main task of the imagination will therefore not be to synthesize or harmonize what is already given as a togetherness in human experience. I have attempted to define the uniqueness of Dilthey's theory of the imagination by assigning it an articulative role.[11] The function of articulation is to bring out certain points of determinacy in a continuum that is initially indeterminate. To articulate the nexus of ordinary consciousness is to explicate its meaning structure. In his essay on the poetic imagination, Dilthey delineates three basic features that define the imagination. The imagination begins by excluding indifferent features of what we find in consciousness and then intensifies what is left. These processes of exclusion and intensification produce quan-

---

10 Kant, *Critique of the Power of Judgment*, 102
11 See Makkreel, Rudolf (1975): *Dilthey: Philosopher of the Human Studies*. Princeton, NJ: Princeton University Press, p. 201

titative changes in our experience. It is a third process of completion that first introduces qualitative changes in our experience by allowing what is focused on to reflect a more general context and to embody some deeply held values.[12]

To make sense of this completing function of the imagination we must again consider the temporality of consciousness. For Kant time was an ideal form within which phenomenal appearances come and go. For Dilthey, however, time is experienced as a real nexus that is historically funded. What is given in the present nexus of consciousness retains aspects from the past. The things we currently perceive are also apperceived as meaningful on the basis of a background structure that sums up not only what we remember from past experience, but also what we have learned from it and hold to be still valuable. Dilthey calls this organizing and guiding memory framework developed by each person "the acquired psychic nexus."[13] This acquired nexus, which always has an apperceptive *regulative* influence on present experience, assumes a creative *constitutive* role in the completing power of the poetic imagination. By drawing on the acquired psychic nexus, the imagination reconstitutes what has come to be focused on in experience by a process of completion through enrichment. What was implicitly apperceived as valuable about life through the acquired psychic nexus is now gathered and compressed into something particular that can be explicitly perceived. The poet's imagination is exemplary in being able to find particular human situations that typify a more general meaning perspective on the world. Now perception is not merely guided by apperception, but comes to imaginatively embody it.

So far we have delineated Kant's formative account of spatial imaging and his representational account of the imagination within the ideal time of consciousness and Dilthey's typifying account of the imagination that embeds it in historical time. What is also needed is an account that integrates these spatial and temporal functions of the imagination and locates them in a worldly context. I will develop this as an orientational theory of the imagination and use the idea of contextualization to clarify the move from an imagination that schematizes prefiguratively to one that schematizes configuratively.

---

12 Dilthey, Wilhelm (1985): *Poetry and Experience, Selected Works (hereafter SW)*, vol. 5. Princeton, NJ: Princeton University Press, pp. 93–106
13 Dilthey, Wilhelm: *Poetry and Experience, SW5*, 96–98.

The Imagination as Seeking Contextual Orientation

Kant's imagining subject perceives things spatially and apperceives them representationally in a formal and ideal time. Dilthey's imagining subject presents and articulates the world in a historical time that encompasses the content of what has been acquired from the past. An orientational theory of imagination must be able to integrate these functions based on what it means to hold a specific place in the world. Although Kant did not develop an orientational theory of the imagination, he does offer some suggestive ideas about orientation as such.

To be oriented to the world is to define my relation to some horizon in terms of a capacity to locate my own place in it. Now I am no longer a disembodied mind that perceives representationally and am more than a recipient of what has been historically acquired. Orientation locates me in the world *both perceptually* and in terms of felt relations. *Perception* provides me with a general visual field, but to be able to relate the *specific* quadrant of the world I see in front of me to the other quadrants that surround me requires a "feeling of a difference in my own subject,"[14] according to Kant. He calls this "the difference between my right and left hands" or more generally, between the two sides of my body. The distinction between right and left, which is an immediate sensory discrimination based on a bodily feeling, is indispensable for providing me a sense of direction as I move into the world. I think, however, that Kant's distinction between left and right needs to be supplemented with the contrast between what is perceived in front of me and what must be imagined as lying behind me. Both axes of reference are necessary to orient me to the so-called four quadrants of the world around me.

Kant relates the felt difference between left and right to the sides of my body, but it can be more closely attached to vision itself by noting that normal vision involves both our left and right eyes. What I see is not a synthesis of two independently given viewpoints but a convergent spectrum in which a focal point is surrounded by a peripheral fringe. Orientation allows me to shift my focus within the field of my vision so I can direct myself as I move into the world.

The original meaning of orientation involves being directed on this earth by the location of the sun as it rises in the morning. Knowing that the morning sun points to what is east of me, I can then also locate things that lie west, south, or

---

14 Kant, Immanuel (1998): "What Does It Mean to Orient Oneself in Thinking?". In:Wood, Allen and George Di Giovanni (trans. and ed.): *Religion within the Boundaries of Mere Reason and Other Writings*. Cambridge: Cambridge University Press, p. 4; Ak 8: 134.

north of me. If I need to find something not directly visible, but know that it is further south, I can at least imagine how I might get there.

Here I am not just oriented by the position of the sun in the sky, but also orient myself toward a desired destination. The orientational imagination does not schematize a concept by anticipating how it applies to some directly available object. Instead it schematizes the path that must be traveled to reach *an* object that is not already in sight but which I nevertheless seek. In charting this path I move from the more familiar to the less familiar. And if I encounter an obstacle I must decide whether I should proceed leftward or rightward to best reach the sought for destination.

Kant uses the experience of spatial orientation in order to then apply it to the problem of orienting ourselves in pure thought. That is, he asks how we should think about the "immeasurable space of the supersensible,"[15] where there are no potentially visible objects to direct us. In such a supersensible world a subjectively felt need of reason can be our only guide. Kant defines this felt need as a faith in reason that must preserve inviolate the freedom to think and to accept no laws "except those which reason gives itself."[16] Here the felt physical difference between left and right is replaced with the felt normative difference between right and wrong.

What is ultimately important about orientation for Kant is that it involves responding to the world on one's own terms. It requires more than being guided from without, namely, a capacity to direct oneself from within. Leaving the supersensible behind, let us return to the world of experience and what can be thought about it. For this world is not just what our senses teach us about it. To flesh out what an orientational imagination can be, we must be able to find a middle ground between sensible perception and pure thought. At this point, we should note that it no longer suffices to speak of the imagination as a special power to produce something distinctive such as imagery or schemata. We can exhibit imagination in how we think about things and judge situations, how we respond and act, and in the ways we express ourselves and become productive. Even the creative imagination of artists who propose new possibilities will not produce great works of art, according to Dilthey, unless they can also "advance our ability to interpret reality."[17] To put this in Kantian terms: the idiosyncratic imagery of a genius must be made communicable to human beings in general.

---

15 Kant, "What Does It Mean to Orient Oneself in Thinking?" p. 6.
16 Kant, "What Does It Mean to Orient Oneself in Thinking?" p. 12.
17 Dilthey, *Poetry and Experience, SW 5*, 211.

To speak of an orientational imagination is shorthand for the capacity to respond to the world in a contextualizing way. It involves the capacity to hold things together in a horizon while also making differentiations within it. This will enable us to eventually shift from the perceptual language of *finding* the placement of things to the hermeneutical language of *seeking* the appropriate contexts for understanding their meaning. But let us stay at the level of placement for a moment. Things are not simply located side by side in mere juxtaposition, but take part in constellations that allow us to discern convergences as well as divergences. When I rest my two hands on a surface in front of me, I can see them to be incongruous counterparts that are pointed in different directions, but when I move my hands to face and touch each other I make them overlap. They disclose themselves as sensory counterparts that exhibit both a divergence in direction and a convergence in scope. When my hands rested on the surface in front of me, I saw them to be incongruous, but as part of my lived body they are felt to be symmetrical. These internal differentiations are orientational and assume the capacity to turn things over to test for potential overlap as we interface them. Similarly, the orientational imagination can be said to turn things over in thought, thus allowing us to compare and contrast them as we shift and rearrange them.

Kant developed a set of comparative concepts that also serve to contrast things. He calls them reflective concepts and offers four pairs of them in the *Critique of Pure Reason*. They are identity vs. difference, agreement vs. opposition, inner vs. outer, and matter vs. form.[18] Since Kant orients us to both the sensible and supersensible levels of reality, his reflective concepts are meant to provide clues for differentiating them. The phenomenal objects of nature whose in-itself content is unknowable can only be approached from without. Thus marks of difference and opposition, and delineations of outer form will take precedence in our cognition of nature. Any theoretical judgments claiming to focus in on the identity, agreement and inner material content of the phenomenal objects of nature will be derivative and hypothetical. By contrast, when we think of noumenal or intelligible objects, considerations of identity, agreement, and inner content must come first according to Kant. Since Kant's duality of the phenomenal and noumenal is no longer pivotal for a contemporary worldly orientation, we can reshape these reflective concepts to be less sharply bivalent. I propose that the identity-difference contrast be transformed into a similarity-distinctness relation, the agreement-opposition contrast into a coordination-juxtaposition relation, the matter-form contrast into a relation between content and its articulation, and finally the inner-outer contrast into a convergence-divergence relation.

---

18 See Kant, "On the Amphiboly of Concepts of Reflection," *Critique of Pure Reason*, A262/B318.

The latter shift is especially important for human experience, since it has become increasingly obvious that any sharp separation of inner and outer is difficult to uphold. Our supposedly inner feelings are usually about things outside us. If we conceive ourselves as worldly beings nothing can be totally external or indifferent to us. Thereby we enter a nexus of relations within which some matter more and others less. What matters most to us can then be defined as the point in this nexus or continuum where divergent interests converge.

Instead of separating levels of reality such as sensible and intelligible objects or two kinds of experience (inner versus outer), the orientational imagination can distinguish multiple worldly spheres to further specify organize human experience. Orientation is often conceived as a relation of a single subject to the world, but this ignores the intersubjective or social dimensions of our experience of the world. As we consider the spheres to which Kant's theory of judgment relates our consciousness we will at the same time look for ways in which they can relate the private and the public.

The cognitive claims of Kant's *Critique of Pure Reason* conceive all objects as part of nature considered as one overarching lawful domain. Accordingly, each object is thought to be determined from without, but this becomes a problem for Kant in the *Critique of the Power of Judgment* where objects are to be judged for their inherent value. In order to make room for the aesthetic appreciation of a specific object, Kant indicates that we must recontextualize it by directing our attention away from the abstract theoretical context of a domain to that of a richer territory of experience.

In the *Introduction* to the *Critique of Judgment*, Kant makes the more general claim that whenever we judge an object we at the same time consider it as part of some context and then proceeds to lay out a total of four such contexts. The logical space that an object occupies merely by virtue of being thinkable is called a *field (Feld)*. The part of this field that is actually experienceable is a *territory (Boden)*. This context can in turn be conceived as a cognitive *domain (Gebiet)* to the extent that it is recognized to be governed by universal laws. But there are also parts of the territory of experience where the order we find is merely based on familiarity and has not been derived from universal laws. Kant calls this kind of context a habitat (*Aufenthalt*) or local sphere. Here we apply empirical concepts to cope with the contingencies of life. What sets a habitat apart is that it also establishes the location of the human subject of the judgment of taste. It provides the familiar and familial point of orientation from which the world is evaluated. Summing up what we have said about these four contexts, we can reconceive them in modal terms and speak of a field of possibilities, a domain of necessity, a territory of the actual and a habitat of contingency. These contexts are not merely distinct spatial regions relative to our earthly location, but provide various ways of ordering the world of our experience.

Let us consider briefly how these distinctions can be applied to our aesthetic enjoyment and understanding of works of art. Since there is always something unexpected about aesthetic pleasure, we can regard our initial relation to appreciated objects as opening up a contingent habitat. Then the task of the imagination is to relate this local and private sphere of the pleasurable, not only to the actual territories of shared taste as manifested in fashion, but also to the field of a possible human agreement of taste about aesthetic values such as beauty and sublimity. Whether the aesthetic imagination also broaches the domain of necessity is more controversial, but certainly the great Greek tragedies evoke it imaginatively through their appeal to fate. In considering this range of modal contexts, the imagination's orientational task is to play them off against each other.

Kant's theory of imagination has been shown to involve two kinds of schematization: the conceptual schematization that prefigures the meaning of objects of experience and the non-conceptual schematization that configures the various contexts to which human subjects can have their feelings attuned. However, in Section 59 of the *Critique of the Power of Judgment*, Kant introduces a third or symbolic function of the imagination that can be regarded as schematizing with ideas. When discussing the symbolic schematization of the imagination Kant himself makes reference to two of the four orientational contexts we have discussed, namely, the territory of experience and the domain of practical reason. Kant is well-known for calling beauty a symbol of morality. In doing so he allows an aesthetic idea about sensible objects to be symbolically related to a rational idea about supersensible freedom. Thereby an abstract non-intuitable rational idea finds an indirect intuitive counterpart in an aesthetic idea. The power of the imagination that makes this possible is sometimes described by Kant as *Gegenbildung.* Conceived literally, *Gegen-bildung* is a process of counter-formation, which is then explicated as a symbolic analogue formation that relates different level contexts.

Although Kant had disciplined discursive philosophy not to expect rational ideas to be intuitively or directly presented, symbolic presentation provides a way for them to obtain indirect intuitive analogues. To be sure, philosophers should not expect the rational idea of God to be intuitively demonstrable and to provide determinate knowledge. But through reflective judgment we can draw on intuitive analogies from the familiar human world to illuminate the field of the transcendent. When it is asserted in the *Critique of the Power of Judg-*

ment that we can have "cognition of God" that is "symbolic,"[19] this will not be knowledge determining the nature of God, but merely a reflective interpretation of our possible relation to God. And Kant's general strategy in this work is to locate certain affinities between the otherwise different domains of nature and freedom by using sensuous beauty to exemplify the moral attributes of human freedom.

In his discussion of beauty as a symbol of morality Kant shows that we commonly ascribe moral qualities to beautiful objects: "We call buildings or trees majestic ...; even colors are called innocent, modest or tender, because they arouse sensations that contain something analogical to the consciousness of a mental state produced by moral judgments."[20] Such morally tinged "aesthetic attributes" lead to the formation of aesthetic ideas which are the felt counterparts of moral ideas. It is these intermediary aesthetic ideas that are symbolically expressed in beautiful forms.

It may seem odd that Kant claims that even natural beauty involves the expression of aesthetic ideas. But he makes it clear that in doing so he is not assigning aesthetic attributes to natural objects in a realistic sense, for aesthetic judgments function according to the principle of the idealism of purposiveness of both nature and art. Thus after asserting the "the song of the bird proclaims joyfulness and contentment with its existence," he adds, "at least so we interpret nature, whether anything of the sort is its intention or not."[21] To regard a bird's song as expressive of joy is not to make a determinant claim about any conscious design in nature, but to make a reflective judgment about a formal purposiveness of nature in relation to our own theoretical and practical ends. This means that the purposiveness in the beauty of nature is a product of interpretation. In judging natural beauty "the important point is not what nature is, or even what it is for us as a purpose, *but how we take it.*"[22] The aesthetic idea provides a rule for the imagination to indirectly present something supersensible in sensible terms. Kant's imagination finds reflective counterparts at the different levels of the sensible and supersensible.

Applied to our orientational imagination, *Gegenbildung* or counter-formation will function on just the one level of what is sensible, thereby granting us the capacity to differentiate the specific worldly contexts that are relevant to us as individuals. Accordingly, a symbolic worldly imagination will aim at analogies that relate things from the various contexts we use to organize and interpret

---

19 Kant, *Critique of the Power of Judgment*, §59, Ak 5: 353.
20 Kant, *Critique of the Power of Judgment*, §59; Ak 5: 354.
21 Kant, *Critique of the Power of Judgment*, §42, 181; Ak 5, 302.
22 Kant, *Critique of the Power of Judgment*, §58, 224; Ak 5: 350, emphasis added.

our experience. Although the *Critique of the Power of Judgment* is best known for suggesting speculative relations between beauty, morality and the divine, it also shows that the symbolic imagination can use our understanding of how things are related in a familiar and intuitable context in order to illuminate how things are related in another less familiar but still worldly context. Thus Kant indicates that the rather abstract contrast between despotic governments ruled by a "single absolute will" and constitutional governments ruled by "laws internal to the people"[23] can be made more intuitively concrete by imagining the former as a machine and the latter as an "animate body." There is no congruence of content here because we are comparing how things function in a natural context to how they function in an ideal political context. Nevertheless, there is a formal analogy that can serve a heuristic function. We can apply "the form of reflection"[24] suggested by the more tangible natural context to the less tangible political one. To make a machine as the symbol for a despotic state is to bring out that its governance is artificially controlled and imposed from without; to imagine a constitutional state as an organism is to see it as being self-regulating and organized from within. The poles of machine and living organism establish the bounds for a whole range of possible political states.

Analogies among different contexts are by their very nature indirect. Organisms and constitutional states cannot be directly compared in terms of content. But the ways they function for us disclose limited formal parallels. Reflection is needed to understand what those limits are as we keep contextual differences in mind. To fail to recognize these differences is to fall into what Kant calls an amphiboly or confusion.[25] Just as amphibious creatures adapt themselves so that they can live both on land and in water, we human beings must learn to respond differently to each of the various contexts that we need to make sense of things.

I consider it to be one of the main tasks of hermeneutics to orient us to these distinctive contexts and thereby combat contextual confusions. The full orientational scope of hermeneutics echoes reason's quest for an overall coherence of things, but in light of the failure of traditional philosophy to establish the subordination of everything to one overarching system, philosophical hermeneutics must seek for a critical coordination of varied frames of reference. Coordination is often thought of as a simple process of juxtaposing things and relating them externally, but the coordinative task of hermeneutics also includes a consideration of the ways different meaning contexts may intersect and at times partially

---

23 Kant, *Critique of the Power of Judgment*, §59, Ak 5: 352.
24 Kant, *Critique of the Power of Judgment*, §59, Ak 5: 351.
25 Kant, *Critique of Pure Reason*, A260/B316

converge. This goes beyond the way Kant allowed symbolism to create parallels between an organism as a teleological system and the ideal way a republic is supposed to function as a political legal domain. When contexts intersect, the forces at work in one may lead processes in the other to be diverted. Actual nation states that are constituted as republics are often seen to deviate from their ideals as we consider their historical development. We need only imagine the way an international economic system intersects with the workings of a nation state and its system of laws to recognize such deviations. Each system has its own forces at work and the balance of power may in turn be conditioned by other factors like the geographical sphere and the material resources that a nation state has available at a specific time. A large country may be able to resist negative market forces and the burdens of international currency exchange that smaller countries with fewer resources cannot. Here the contextual configurations of the orientational imagination must incorporate temporal as well as spatial vectors.

We started with a simple perceptual imagination that produces a spatial synopsis and moved on to imaginative syntheses to temporally order experience in accordance with the laws of causality. Then we considered how the imagination schematizes either prefiguratively to produce meaning content or configuratively to orient us contextually. Finally, we saw how Kant's account of the symbolic imagination finds reflective analogies or parallels among distinct contexts. But for hermeneutics, which must reflect on the way various contexts intersect, an articulative conception of the imagination will prove more useful than one defined as synthetic. As we accumulate experience the nexus of the meaning of our life becomes ever more determinate. The function of imaginative articulation is to gather the various threads of our experiential input at points where relevant meaning contexts can also be thought to intersect. When this succeeds, then the orientational imagination secures us our place in the world and our part in history. Ultimately, the imagination has a historical task in rounding out our experience.

The Imagination as Finding and Configuring Meaning Contexts

Kant proposed four kinds of context for human judgments. However, there is no reason to not also leave room for other contexts such as the more specific *systems* that the scientific disciplines have carved out for themselves within the philosophical domains of theoretical and practical reason. By correlating systems with disciplines, we can generate functional contexts capable of interacting to some extent. Several contexts can frame the same object–we saw Kant consider objects in light of their territorial location as well as in terms of the laws of a domain. Similarly, various social and cultural systems can converge on the same historical subject-matter. As part of its orientational task, reflective judgment must consider questions of priority and decide which referential context takes precedence. Otherwise, confusions and misunderstandings will arise just as Kant warned in his discussions of reflective amphibolies. While it is important to prioritize subordinative relations of dependence within a legislative domain, it makes more sense to focus on coordinative relations of interdependence when exploring the territories of human exchange and communication.

In light of all this, *what* the imagination seeks must also specify a circumstantial *where*. The meaning contexts we choose to bring to bear must be properly correlated so that one standpoint does not close us off from another. This requires setting the right priorities so that actual dependence relations are not misunderstood or distorted. We must always be aware of the limits of each kind of contextual perspective while at the same time remaining oriented to the world at large. If we fail to perform this balancing act we can fall into prejudices. And when we allow the various relevant contexts of our experience to randomly merge into each other, we live as in a daydream in which reality may become illusory. The more powerful delusions of our night dreams seem to derive mostly from a lack of worldly orientation. What we imagine while dreaming often seems to come out of nowhere. Either we neglect to specify the proper context of what we see happening or we have merged several contexts leading to disorientation. The pleasures we draw from certain dreams may very well stem from either indeterminate or confused contextualization. Conversely, the claustrophobic oppressiveness of a nightmare bespeaks the dominance of a very restrictive context. What a relief it is to wake and again find our everyday bearings.

Kant speculated that we need to dream while asleep so that our imagination will keep our "vital organs" going while we are closed off from experiential stimuli. He thought that dreams exert "an internal motive force" without which "sleep ... might well amount to a complete extinction of life."[26] Regardless of whether the imagination has this physical function, it is obvious that it can have a therapeutic mental effect while we dream. It allows us to escape the emotional burden of coping with the incessant demands of life as well as to suspend

---

26 Kant, *Critique of the Power of Judgment*, §67, Ak 5: 380.

the reflective task of choosing the most appropriate contexts for assessing and interpreting our experience.

What then is the orientational imagination? Can any of the more traditional mediating functions of imaginative synthesis and harmonization still be salvaged? They can to the extent that we focus on what is perceived and felt and assume that they can be referred to a coherent world. For Kant this world was first of all the phenomenal domain of nature. Dilthey enlarged this world to include the fundedness of historical life. For him the poet's imagination takes the normal everyday context of some important event and completes it on the basis of his or her acquired psychic nexus to articulate some more general perspective on life as such. But to the extent that the imagination seeks the meaning of things in today's multicultural world, it must be able to adjust itself to quite disparate contexts. Then the artist may also illuminate his or her everyday context by interfacing it with some more remote contexts that can raise questions about our basic assumptions. The orientational function of the imagination will be to navigate among different contextual positions just as it was its initial schematizing function to modulate from the medium of conceptual thought to the medium of sense. But the additional hermeneutical task of imaginative orientation is to take the measure of the various intersecting contexts and to rank their appropriateness for what is to be understood.

Index

www.ingramcontent.com/pod-product-compliance
Lightning Source LLC
Chambersburg PA
CBHW070031100426
42740CB00013B/2661